AF841887

East Asia's Demand for Energy, Minerals and Food

China, Japan and South Korea's international relations are shaped by the fact that all three countries are significant importers of resources. This book brings together work on specific aspects of the politics of resources for each of these countries, regionally and internationally. There are some similarities in the approaches taken by all these three. For example, their development assistance shares a focus on infrastructure building and reluctance to purposefully influence domestic politics. However, there are also significant differences due in large part to the individual nature of the states as international actors. China has significant domestic supplies of resources while Japan and Korea are net importers. China's size also marks it out as different, as does its state socialist history and continuing authoritarian state. One of the key issues to understanding contemporary resource politics in Northeast Asia is that Western dominance of the world order is currently declining. In some cases Northeast Asian approaches to resources are seen as being mercantilist. In other cases Northeast Asian powers are seen as replacing Western powers in exploiting resource-rich developing countries. This book gives readers an informed view of this very important issue in contemporary international relations.

This book was previously published as a special issue of *Asian Studies Review*.

Kate Barclay is a senior lecturer of the University of Technology, Sydney, Australia. She researches the social aspects of the production and trade of food, especially fisheries in the Asia Pacific region. One current research interest is the sustainable development of tuna resources in the context of changing governance systems and globalisation.

Graeme Smith is a senior research fellow at the State, Society and Governance Program, Australian National University, Australia. His research has explored the politics of agriculture in China and Chinese aid, investment and migration in the Asia-Pacific region.

East Asia's Demand for Energy, Minerals and Food

The International Politics of Resources

Edited by

Kate Barclay and Graeme Smith

Routledge
Taylor & Francis Group
LONDON AND NEW YORK

First published 2015
by Routledge
2 Park Square, Milton Park, Abingdon, Oxon, OX14 4RN, UK

and by Routledge
711 Third Avenue, New York, NY 10017, USA

Routledge is an imprint of the Taylor & Francis Group, an informa business

© 2015 Asian Studies Association of Australia

All rights reserved. No part of this book may be reprinted or reproduced or utilised in any form or by any electronic, mechanical, or other means, now known or hereafter invented, including photocopying and recording, or in any information storage or retrieval system, without permission in writing from the publishers.

Trademark notice: Product or corporate names may be trademarks or registered trademarks, and are used only for identification and explanation without intent to infringe.

British Library Cataloguing in Publication Data
A catalogue record for this book is available from the British Library

ISBN: 978-1-138-79631-7

Typeset in Times New Roman
by RefineCatch Limited, Bungay, Suffolk

Publisher's Note
The publisher accepts responsibility for any inconsistencies that may have arisen during the conversion of this book from journal articles to book chapters, namely the possible inclusion of journal terminology.

Disclaimer
Every effort has been made to contact copyright holders for their permission to reprint material in this book. The publishers would be grateful to hear from any copyright holder who is not here acknowledged and will undertake to rectify any errors or omissions in future editions of this book.

Contents

Citation Information

The chapters in this book were originally published in *Asian Studies Review*, volume 37, issue 2 (June 2013). When citing this material, please use the original page numbering for each article, as follows:

Chapter 1
Introduction: The International Politics of Resources
Kate Barclay and Graeme Smith
Asian Studies Review, volume 37, issue 2 (June 2013) pp. 125–140

Chapter 2
China and Japan in Myanmar: Aid, Natural Resources and Influence
James Reilly
Asian Studies Review, volume 37, issue 2 (June 2013) pp. 141–157

Chapter 3
Chinese Aid in the South Pacific: Linked to Resources?
Philippa Brant
Asian Studies Review, volume 37, issue 2 (June 2013) pp. 158–177

Chapter 4
Nupela Masta? Local and Expatriate Labour in a Chinese-Run Nickel Mine in Papua New Guinea
Graeme Smith
Asian Studies Review, volume 37, issue 2 (June 2013) pp. 178–195

Chapter 5
China, Natural Resources, Sovereignty and International Law
Ben Saul
Asian Studies Review, volume 37, issue 2 (June 2013) pp. 196–214

Chapter 6
Securing Fish for the Nation: Food Security and Governmentality in Japan
Kate Barclay and Charlotte Epstein
Asian Studies Review, volume 37, issue 2 (June 2013) pp. 215–233

Chapter 7
Rare Earths: Future Elements of Conflict in Asia?
Ming Hwa Ting and John Seaman
Asian Studies Review, volume 37, issue 2 (June 2013) pp. 234–252

Chapter 8
Throwing the Baby Out with the Bathwater: Australia's New Policy on Treaty-Based Investor-State Arbitration and its Impact in Asia
Luke Nottage
Asian Studies Review, volume 37, issue 2 (June 2013) pp. 253–272

Please direct any queries you may have about the citations to clsuk.permissions@cengage.com

Notes on Contributors

Kate Barclay researches the social aspects of the production and trade of food, especially fisheries in the Asia Pacific region. One current research interest is the sustainable development of tuna resources in the context of changing governance systems and globalisation, especially the opportunities and pitfalls presented by "ethical consumption". Journals in which her work has appeared include *Critical Asian Studies*, *Japan Forum*, *The Contemporary Pacific* and *Marine Policy*. Her books include *Foreign Bodies in Tinned Tuna* (2008), *Capturing Wealth from Tuna* (2007) and *Engaging with Capitalism: Cases from Oceania* (2013). Kate is a Senior Lecturer in the International Studies program at the University of Technology, Sydney, Australia.

Philippa Brant is a Research Associate at the Lowy Institute for International Policy. She completed her PhD at The University of Melbourne, Australia, in 2012 where she investigated China's foreign aid program and the implications for the global aid architecture. Her recent publications include *South–South Cooperation: A Concept Paper* (International Poverty Reduction Centre in China, 2011). She has been a recent panelist with Asialink, the Australian Centre on China in the World, and publishes commentary pieces and blog posts with *The Lowy Interpreter* and *whydev.org*. Her current interests cover Chinese foreign policy, Chinese engagement in Southeast Asia and the Pacific, China's impact on global governance norms, and Australia–China relations.

Charlotte Epstein is an Associate Professor at The University of Sydney, Australia. Prior to that she was a Georges Lurcy Visiting Scholar at the UC Berkeley Political Science Department, USA. She is currently visiting professor at the Centre d'Etudes et de Relations Internationales (Paris) and chair of the International Studies Association's International Political Sociology section. Her main interests are the international politics of natural resources and the role of language in international politics. She is the author of *The Power of Words in International Relations* (runner up to the 2009 ISA Sprout Award), and has published in *Global Environmental Politics*, *European Journal of International Relations* and *International Political Sociology*. She has a forthcoming article in *International Organization*.

Luke Nottage specialises in arbitration, contract law and consumer product safety law, with a particular interest in Japan and the Asia Pacific. He is Associate Dean (International) and Professor of Comparative and Transnational Business Law at Sydney Law School, founding Co-Director of the Australian Network for Japanese Law, Associate Director of the Centre for Asian and Pacific Law at The University of Sydney, and Comparative and Global Law Program coordinator for the Sydney Centre for International Law. His publications include *Corporate Governance in the 21st Century: Japan's Gradual Transformation* (2008), *International Arbitration in Australia* (2010), *Foreign Investment and Dispute Resolution Law and Practice in Asia* (2011), several other books, and more than a hundred chapters and refereed articles.

James Reilly is a Senior Lecturer in Northeast Asian Politics at The University of Sydney, Australia. He is the author of *Strong Society, Smart State: The Rise of Public Opinion in China's Japan Policy* (2012) and co-editor of *Australia and China at 40* (2012). His articles have appeared in journals including *Asian Survey, China: An International Journal*, *Chinese Journal of International Politics*, *Japanese Journal of Political Science*, *Journal of Contemporary China*, *Modern Asian Studies*, *Survival* and *Washington Quarterly*. He has also published several book chapters in edited volumes. He holds a PhD in Political Science (2008) and an MA in East Asia Area Studies (1999). He served as the East Asia Representative of the American Friends Service Committee (AFSC) in China from 2001–08.

Ben Saul is Professor of International Law and Australian Research Council Future Fellow at The University of Sydney, Australia. He is internationally recognised as a leading expert on global counter-terrorism law, human rights, the law of war, and international crimes. He has published six books, 65 scholarly articles and hundreds of other publications and presentations, and his research has been used in various national and international courts. Ben has taught law at Oxford, the Hague Academy of International Law, and in China, India, Nepal and Cambodia, and has been a visiting professor at Harvard Law School. He practises as a barrister in international and national courts, has advised various United Nations bodies and foreign governments, has delivered foreign aid projects, and often appears in the media.

John Seaman is a Research Fellow in the Center for Asian Studies of the French Institute of International Relations (Ifri), where he specialises in Chinese energy and foreign policies, questions of energy security and geopolitics in Asia and policy approaches to critical raw materials issues worldwide (including rare earth elements). He earned Bachelor's degrees from Seattle University, USA (International Economics; French) and a Master's degree from Sciences Po Paris (International Affairs – International Security). In 2002–03 he was a NSEP David L. Boren Scholar at the Beijing Center for Chinese Studies. He is also currently an International Research Fellow at the Canon Institute for Global Studies (CIGS) in Tokyo. He has spent a number of years studying and working in both China and Japan.

Graeme Smith is a Senior Research Fellow at the State Society and Governance in Melanesia program at the Australian National University, Australia. His research has explored the demand for organic produce in Chinese urban centres, the political economy of service delivery in rural China, and the redistribution of land among Chinese farmers. He also studies Chinese investment in the Asia-Pacific region, with ongoing projects in Papua New Guinea, New Caledonia and Myanmar. His journal articles have appeared in

The China Journal, *Pacific Affairs* and *The Journal of Peasant Studies*. He also holds a PhD in environmental chemistry, has written several guidebooks to China, and is the 2011 winner of the Gordon White Prize for the best article published in *China Quarterly*.

Ming Hwa Ting works for the South Australian Government. His work has been published in journals including *Global Asia*, *New Zealand Journal of Asian Studies* and the *Taiwan Journal of Democratization*. His most recent research on the global demand for uranium will be appearing in a forthcoming edited volume published by ABC-Clio. He is now researching the dynamics of Chinese and Middle Eastern investment in Africa.

Introduction: The International Politics of Resources

KATE BARCLAY

University of Technology Sydney

GRAEME SMITH

The University of Sydney

Abstract: *China, Japan and Korea's international relations are shaped by the fact that all three are significant importers of resources. This Introduction proposes two conceptual frameworks for understanding the politics that is taken up in the papers of this Special Issue. The first is to consider the extent to which there is an East Asian model of resource procurement. We find that there are some similarities in the approaches taken by all three countries; for example, their development assistance shares a focus on infrastructure building and a reticence to purposefully influence domestic politics. There are, however, also significant differences due in large part to the individual nature of the states as international actors. The second conceptual framework is the broad contemporary theme of the end of Western dominance of the world order. The main way this affects the international politics of resources in Northeast Asia is through the belief that the activities of those countries are threatening in some way. In some cases Northeast Asian approaches to resources are seen as a problem because they are not sufficiently liberal, whereas in others the problem is that Northeast Asian powers are seen as replacing Western powers in exploiting resource-rich developing countries.*

Introduction

Japan is well known as a resource-poor country, while China has come to be seen as resource-hungry over the last decade or so. South Korea's demand for resources is of a lower order but is nonetheless significant; it is among the world's top importers of energy, and also imports large proportions of the minerals and food it uses. Securing access to resources has been a key driver in Japan's foreign policy since Japan opened up and embarked on the road to modernisation in the late nineteenth century. Following the example of European powers, Japan used colonial and military strategies up to 1945. Since then, Japan has used various forms of diplomacy, including development assistance and free trade agreements, to facilitate its access to energy, minerals and food via international trade. The government encouraged the private sector to invest in resources production overseas to expand supplies of resources. Japan has also protected its domestic food-producing sector. Korea's strategies for securing resource supplies have centred on diversifying supply as well as using taxes, tariffs and industrial restructuring to reduce vulnerability to world energy markets. Like Japan, Korea protects against the effects of food imports on the rural economy. China has relatively greater domestic supplies of resources, but as its economy has grown so has its reliance on resources from elsewhere. Its efforts to secure supplies of resources from overseas are often portrayed as predatory. China, being a net food exporter, has not intervened in food markets in the same way as Japan and Korea, but its restrictions on exports of rare earths used in global high-tech industries show China's willingness to implement trade barriers in some circumstances.

The resources factor is significant in East Asian countries' relations with Africa, Latin America and the Asia Pacific to the point where it may be framed as a security issue; either energy security or food security. This reality raises questions about patterns of development assistance, notably about the extent to which aid diplomacy might be connected to the resource offerings of recipient countries. Japan has been a significant power in world politics for more than a century, China is rapidly becoming a major power, and Korea has settled into position as a solid middle power. These are not insignificant international leadership roles, so it is important to consider how the politics of resources in these countries plays out both regionally and globally.

In bringing together scholars looking at the politics of resources in East Asia, this Special Issue offers a comparative perspective on the similarities and differences between the approaches taken by each country to secure access to crucial energy, mineral and food resources, and the significance of these efforts in terms of international politics. One main question considered by papers in the collection is the extent to which there is an identifiable East Asian approach to resource investment. We do not find a neat "model" suggesting wholesale emulation of first-mover Japan's strategies by Korean and Chinese actors; however, we do find some similarities. These include: 1) assistance to develop infrastructure for resource extraction/production and transport in resource-exporting countries; 2) a "win-win" approach overlaying purported altruism in Official Development Assistance (ODA) associated with resources; 3) foreign direct investment (FDI) working in collaboration with ODA to facilitate resource exports from developing countries; and 4) policies that attempt to delink ODA from political influence in the recipient country. There are variations, however, in the extent to which the above have been applied by each government, and also considerable differences in the ways in which different corporations from each of the East Asian countries have conducted their investments in the mining, energy and food sectors. Variations in approach are caused by experiences of individual investors in host countries, and processes shaping the approaches of policymakers in host and investor countries. Despite the similarities noted above, therefore, it is equally accurate to say that the East Asian approach to resource investment is akin to Deng Xiaoping's maxim of "crossing the river while groping for stones".

Another theme explored in this Special Issue is how world historical trends affect the ways in which East Asian resource investment occurs, as well as understandings of that investment. When East Asian investment in resources is regarded as threatening, the potential for conflict increases, and the rise to prominence of Asian powers immediately upon the end of 500 years of Western hegemony has contributed to the growth of such perceptions. One discourse of the "China threat" may be seen as a reaction by Westerners – used to their own companies and governments being dominant – to the rise of a non-Western power. Another discourse of the China threat arises from developing countries, many of which have experienced centuries of domination by the West, being concerned about yet another power exploiting them.

An East Asian Approach to Resource Investment?

There are some similarities in the approaches taken by East Asian countries to resource development. Regardless of whether such similarities bear close scrutiny, it is important to note that it is widely perceived that such a model exists, and that it is seen to be different from the "Western" model.[1] In the 1980s, this observation was framed in the concept of the East Asian "developmental state", which was juxtaposed with the Western "regulatory state" (Johnson, 1982). More recently it has been presented as the "Beijing Consensus" (Ramo, 2004; Kennedy, 2010a) rising in opposition to the "Washington Consensus" (Williamson, 1990).

As a starting point in searching for a distinctive East Asian approach, we might venture to suggest that there is a shared emphasis on maintaining a politically neutral stance in overseas resources development: the oft-stated principle of "non interference" in the affairs of the host country. In the 1950s, Japan's diplomacy operated under the "Yoshida Doctrine" whereby Japan attempted to stay out of politics and concentrated only on economics as a way to avoid a backlash due to its war record (Tarte, 1998). By contrast, Western governments have preferred to be politically interventionist with their development assistance, withholding it (albeit selectively) from governments they have seen as illegitimate. In the post-Cold War world this policy has been called the "Washington Consensus", by which Western governments have used their aid to encourage policy prescriptions they associate with "good governance" in developing countries. During the 1990s, Japan was pressured to abandon its reticence to intervene politically and to join the Washington Consensus (Reilly, 2012). Whether or not Ramo (2007) intended it to,[2] his term "Beijing Consensus" has become shorthand for China's policy of offering development assistance, trade and investment free of political preconditions, and expecting that each country will follow a distinct path of development. As Shaun Breslin points out, the Beijing Consensus, or the "China model", has come to stand for what China does not represent, rather than what it does. No shock therapy, no inevitable democratisation, no relinquishing of state control, not Western, not neoliberal, not being told what to do by others, not telling others what to do (Breslin, 2011, pp. 1338–39). Neither Japan nor South Korea is considered to be part of the Beijing Consensus, but all three countries certainly share a distinctively strong respect for the principle of inviolable sovereignty that inclines them in a similar direction.

Several papers in this issue point out that there are limits to the policy of non-interference, as well as a growing critique, even within China, of whether such an approach is wise. In the wake of the recent decision by the Myanmar regime to bow to popular pressure and suspend construction of the massive Myitsone Dam, Zhu Feng, the deputy director of the Center for International and Strategic Studies at Peking University, noted:

> The dam's Chinese investors, for their part, relied too heavily on the depth of the two countries' bilateral ties, and so heavily discounted the project's political risks. Their behaviour also reflects the implied guarantee of official government mercantilism, as well as the complacency of China's state-owned enterprises, which account for most Chinese overseas investment. Operating on the assumption that the government will back them – or bail them out if they fail – they can afford to be cavalier (Zhu, 2011).

The question of whether East Asian strategies are liberal or mercantilist is another influential factor in imaginaries of resource investment. In this debate liberal strategies prefer resources to be accessed through private sector international trade and cross-border investment, and issues of national concern to be negotiated and protected through the mediation of international agencies. By contrast, strategies labelled mercantilist are more state-centric and robust, and tend to reach reflexively towards more direct measures, such as boosting military/naval capacity to defend supplies of resources, government support of trade and investment to secure resource supplies, and non-participation in relevant international regimes.

Accusations of mercantilism levied at East Asian states often contain the implication that (Western) liberal approaches to securing resources through international agreements and trade are better in terms of international peace and economic growth (Barclay and Epstein, this volume). Chinese, Japanese and Korean actors do, however, generally abide by international regimes (for China see Saul, this volume), and the economies of all three are heavily dependent on world trade. As John Ikenberry (2010) has pointed out, while it is the case that the Western-led nature of the world order is currently changing and the world order will not be Western-led in future, all the signs are that the liberal nature of the world order is not in decline. He argues that an international system based on open and rules-based cooperation is squarely within the interests of rising Asian powers.

In considering the interests of Northeast Asian governments in this area it is useful to refer to widely accepted definitions of resource security. The United Nations Development Programme World Energy Assessment defined energy security as "the availability of energy at all times in various forms, in sufficient quantities, and at affordable prices" (UNDP, 2000, p. 11). The World Food Summit defined food security as a condition wherein "all people at all times, have physical and economic access to sufficient safe and nutritious food to meet their dietary needs and food preferences for a healthy and active life" (Pinstrup-Andersen, 2009). For both of these definitions security means access. Insofar as world markets and other international regimes provide access, it is then reasonable to assume that states will participate. Markets and international cooperation are both subject to failures, however, so it is equally reasonable to expect that states will hedge their bets with government policies to optimise their benefits from markets (Barclay and Epstein, this volume).

The question may then be posed: are East Asian countries "really" being liberal when participating in international regimes, or are they just being instrumentalist in their use of international regimes to secure access to resources? One of the things to note in this line of questioning is the political significance of being understood as "liberal" or not, as China's supposed mercantilism is related to perceptions of the China "threat". Moreover, the distinction implies that *real* liberal behaviour is motivated by interest in the collective good rather than self-interest. This is a highly questionable proposition, both in terms of liberal understandings of the role of self-interest and in terms of the imputed motivations behind Chinese versus Western actions.

Another useful alternative to the liberal-mercantilist divide is to look at the problem in terms of a given actor's effectiveness in engaging with international norms and laws, in learning how to play the game to serve their own ends (Kennedy, 2010b). In any case, increased concerns about state intervention in securing resources to an extent reflect historical trends. Cold War imperatives drove concerns about resource security in the 1960s and 1970s, leading to ODA being tightly linked to resource diplomacy, and confining the flow of resources to well-defined political blocs (Buijs and Sievers, 2011). The re-emergence of state engagement in the resources sector in part reflects the end of the belief, prevalent in the globalised and liberalised trading environment of the late 1980s and 1990s, that variously-headquartered multinational companies could obtain access to any resources through markets alone.

Furthermore, while there is state involvement in East Asian resource investment, it does not follow that East Asian states are therefore able to exert their national policies through this investment. Many Chinese resource enterprises are state-owned. Much Japanese overseas resource investment has happened in close affiliation with diplomatic efforts, and its domestic food production is protected (Barclay and Epstein, this volume). Until recently most Korean overseas investment in resources was private, but there has been government structuring of domestic markets to shore up resource security, and recently diplomacy is being used to support overseas investment too. When Chinese resource companies face the choice of national interest or commercial gain, the latter nearly always wins out (Rosen and Hanemann, 2009). Disjuncture between the interests of these companies and their industry lobby groups, and the broader strategic and economic goals of their governments is increasingly apparent (Brant, this volume). Recognising their inability to have direct or effective control over investors (Kim, 2008), Korea has been active in drafting regulations to ensure that proceeds from state-supported overseas investment are repatriated and reinvested.

Recipient countries also have agency in these processes, and may thwart the intentions of investor states. For example, Australia has recently gone against an emerging trend in Asia, by announcing in 2011 that it will no longer accept investor-state dispute settlement (ISDS) provisions in future investment treaties. This marks a break with more than two decades of treaty practice, whereby Australia had hitherto consistently obtained ISDS protections in treaties with developing countries with less reliable domestic laws protecting foreign investments. Part of Australia's new concern about ISDS is that developed countries may also be subject to arbitration claims brought directly by foreign investors, alleging illegal interference with their investments – epitomised by a claim brought by a tobacco company in 2011 regarding Australia's plain packaging legislation (Nottage, 2013).

In the short term, Australia's new policy position may not have major repercussions. After all, existing treaties (such as AANZFTA) are not being terminated or renegotiated, so outbound investors in many developing countries in Asia retain ISDS protections. In addition, negotiating partners in bilateral FTAs such as Japan, Korea and China may be prepared to break with their own recent treaty practice and forgo ISDS, in order to secure substantive market access – particularly to Australia's resources sector. In the medium to long term, however, Australia's policy shift may impede conclusion of regional treaties such as an expanded Trans Pacific Partnership (TPP). Furthermore, if states in Asia and elsewhere are convinced by the theory and evidence presented by the Australian Government for the policy shift, it may also lead to the unravelling of the treaty-based ISDS system that has built up world-wide since the 1960s, generating complex implications for cross-border investment flows (Nottage, this volume).

Differences between the three Northeast Asian countries upon which the articles in this Special Issue focus are visible in the nature of connections between government and industry. The existence of authoritarian government in China does not equate with well-organised channels for achieving policy through economic actors. Routes for business groups to lobby governments are highly evolved in Japan and Korea, with large conglomerates and industry groups having extensive influence over government ministries, and vice versa (Barclay and Epstein, this volume; Kim, 2008). Most Chinese enterprises, however, need to rely on informal channels to reach policymakers, much as Indian and Brazilian companies do (Deng and Kennedy, 2010). The exceptions to this general rule are the largest state-owned enterprises, such as the three Chinese oil majors, whose heads are routinely rotated out to ministry-level positions (Jiang and Sinton, 2011; McCarthy, 2013).

Similarities between the approaches of East Asian countries in resource investment lie in conceptualising the provision of ODA in resource-exporting countries as an opportunity for both the recipient and the donor; a "win-win" situation. ODA is linked to an expansion of future investment, trade and resource extraction/production, expanding the capacity of the host country to absorb further investment by East Asian companies, achieved through the development of hard infrastructure (such as roads, railways, ports and dams) and to a lesser extent in health and education. Interestingly, Reilly (2012) argues that China's adoption of this approach to ODA was shaped by its experiences as a recipient of Japanese aid. In this case, Korea seems to present as an exception, as the Korean Government has not had an expansive ODA program (typically around 0.1 per cent of GNI). South Korean corporations have invested in resources without complementary aid/infrastructure support. However, a massive increase in Korean aid to Africa and North Korea will take place as South Korea has committed itself to the Millennium Development Goals of spending 0.25 per cent of GNI on development assistance by 2015 (Ministry of Foreign Affairs and Trade, n.d.). It will be interesting to see whether Korea's aid program will be more like Japan's and China's, with their emphasis on infrastructure and production, or more like those of Western aid donors, who in recent decades have invested heavily in "governance", or take some other focus.

As well as investment connected to ODA, there is straight investment in overseas production to increase the overall world supply of the resources in question. Japan has long followed this policy with regard to food supplies (Barclay and Epstein, this volume). The Chinese Government counters criticism of Chinese investment in energy production in areas such as Sudan by saying it is important to expand overall international supplies of energy.

The preceding paragraphs lay out some broad-brush similarities between the approaches taken by China, Japan and Korea to securing their supplies of energy, minerals and food. There are clearly, however, many important differences. China's vast size, authoritarian polity and history as a socialist command economy seem to make it an outlier in looking for a common "East Asian" approach to resource development, but the impact of these factors on China's approach to resource development is not straightforward. Japan, Korea and China all initially pursued economic development with the aid of a strong state. Moreover, the Chinese state, while authoritarian and likely to remain so for some time, is not necessarily strong or responsive to the wishes of state-owned companies. Size does not equal strength, Mao's dictum notwithstanding. A recent government White Paper entitled *China's Peaceful Development* (the term "peaceful rise" having been shelved because it could be construed to be threatening) plaintively states, "For China, the most populous developing country, to run itself well is the most important fulfilment of its international responsibility".[3] Additionally, while all three countries have more or less combined public and private sector engagements with resource-rich developing countries, they are not unique in this. Indeed, they may be put in the same basket as the European Union, which since decolonisation has arguably pursued a policy of mixing government aid with commercial trade to ensure continued access to resources in former colonies in Africa, the Caribbean and the Pacific, under the Yaounde and Lomé Conventions (Grilli, 1996; Lister, 1988; Ravenhill, 1985).

One notable difference between the three countries' approaches is in policies regarding food production and trade. Food production is a qualitatively different kind of economic activity from that of sourcing minerals and energy. Food production is usually carried out by many small actors, and underpins the economies of rural communities. In most countries most food produced domestically is consumed domestically. Mining and energy industries, on the other hand, have huge investment costs so there are relatively fewer actors, and they are mostly export oriented. Most of the inputs (such as investment, equipment, labour and technical expertise) also come from outside, rather than being articulated throughout local communities such as agriculture and fishing. Including discussion of food as a resource along with minerals and energy as we have in this Special Issue enables us to draw comparisons in the international politics of these sectors. Japan and Korea have both implemented trade barriers to protect domestic food production as a way of handling the social instability that arises with income inequality between urban and rural areas with modernisation (Barclay and Epstein, this volume). China has handled the urban-rural divide very differently, allowing a large income inequality to develop. Also, as a net food exporter, China's outlook on food security is unlike that of the other two, each of which imports well over half of its food supplies. All three, however, share a common historical concern with attempting to maintain food self-sufficiency, particularly in grain.

The End of Western Dominance of the International Order

One of the defining features of world politics as it is emerging in the twenty-first century is the end of the 500 years of domination by the West over Asia, the start of which was marked by the voyages of Vasco da Gama, Magellan and Columbus, and the era of European colonialism that followed (Bell, 2007). During much of the twentieth century Japan was a world power, and now China and India are assuming that mantle in the twenty-first century, while other non-Western states such as Korea are swelling the ranks of middle powers.

The historical juncture of being at the end of 500 years of world domination by the West is a pertinent lens through which to view the international politics of resources. It is important to note that colonialism did not simply disappear with decolonisation but that patterns established by colonial exploitation have continued to be a major influence in the postcolonial world. Legacies of those centuries of Western domination will continue to influence world politics even as Western dominance diminishes. For example, most sources of supply of important energy and mining resources were cornered by Western companies during the period Western powers had dominance. Contemporary Asian investment in energy and mining resources, therefore, is shaped by having to "take what's left", which is mostly in developing countries, and sometimes in countries with records of human rights abuse (Smith, 2012). China and Japan as energy investors have had to establish themselves in a very different environment, for example, in having to deal with state-owned enterprises in independent countries such as Saudi Arabia.

A major way in which the rise of Asian powers affects East Asian resource investment is through sensitivities about that investment, especially through opinions that such investment is threatening in some way. A competitive response on the part of Westerners to East Asian states and corporations joining them in world dominance may explain some of the vociferous denunciations of Asian investment by Western commentators. For their part, former colonies' understandings of Asian investors as neo-colonial are also likely to be coloured by the centuries of plain colonialism by European powers and their investors; the increase in Asian investment may seem like replacing one set of exploitative outsiders with another.

Criticisms of East Asian resource investment

Since the 1960s, Japanese capital has been invested internationally, with the 1980s marking a spike as Japan's economy rose to be the second largest in the world, in the wake of its excess liquidity problems and the need for offshore production. Since the 1990s, the rise of China has eclipsed interest in Japan's demand for resources. These days, the resource issue with which Japan is most commonly associated is overfishing; though concerns about overfishing to meet demand in China are also firmly on the radar. Japan is seen as leading the world now in "guzzling fish", but China, with its massive population eating more protein as it becomes wealthier, is seen as the country to watch in terms of fish imports in the future (Renton, 2005). The concern that China's huge demand for food might lead to global food shortages has been around for a while, with books such as *Who Will Feed China? Wake-up Call for a Small Planet* (Brown, 1995). The discourse around China's overseas resource investment in the English-language media feeds these concerns. For example, China has been described as being on a "relentless mission to capture resources" (McKibbin et al., 2010), spreading "neo-colonial slavery" in developing countries (Callick, 2007) and promoting a development model that is "the most serious challenge that liberal democracy has faced since fascism in the 1930s" (Buruma, 2008).

Japan and China have both been criticised for going against the US- and EU-led international community's avoidance of dealing with "rogue" states (Lee, 2012), as noted earlier in relation to the principle of political non-interference. Considerations about dealing with rogue states may be seen as part of a wider set of questions about the effects on producer countries of trade, investment and development projects in resource sectors. There has been speculation about whether East Asian companies' ways of doing business exacerbate problems of patron-client style governance, causing or contributing to corruption. Japanese trading houses attracted this criticism in the latter half of the twentieth century by working with Southeast Asian governments to secure tropical timber for use in the Japanese construction industry. The investment style led to the development of structures in the logging industries that were ecologically disastrous and damaging to indigenous forest dwellers (Dauvergne, 1997; Tsing, 2005). In the case of China, links have been drawn between Chinese investment strategies and socio-political instability in producer countries. Commentators write of Chinese companies bringing in Chinese labour, or contracting locals under poor conditions, such that great resentment builds, erupting in anti-Chinese riots (Windybank, 2005; Hanson, 2009; McKibbin et al., 2010). In recent years, riots that have included both some kind of anti-Chinese element and a connection with resource investment have occurred in Tonga, Solomon Islands and Papua New Guinea (PNG) (Wesley-Smith and Porter, 2010; Smith, 2012).[4] Chinese-owned mining interests in Zambia are pursuing short-term strategies rather than addressing local environmental and social concerns, and have seen wildcat strikes in response to "precipitous casualisation" of the local workforce (Lee, 2009; Haglund, 2009).

While there are no doubt problems with Chinese ODA and private sector investment in resource extraction globally, it seems that some of the representations of wrongdoing on the part of China are overblown, and contain unexamined assumptions of Western investment as superior. Chinese investment in Africa is often denounced as corrupting and exploitative (see, for example, Mbaye, 2011), but Chinese involvement in Africa arguably follows the pattern of the capitalist world system and is very similar to (not worse than) ongoing Western involvement (Sautman and Yan, 2008). Would a Chinese oil company really be any less ethical than a nominally Dutch company, Shell, has been in Nigeria? Similarly, Chinese aid is often assumed to be morally inferior because it is not used to pursue a "governance" agenda, and to be less effective in achieving "development" than Western aid. Empirical analysis of Chinese aid in Africa, however, shows that much Western aid, particularly that focused on governance, has achieved little, while some Chinese infrastructure aid has achieved a great deal (Sautman and Yan, 2008; Bräutigam, 2010). Chinese and Western donors both pursue their own interests and political discourses in Africa, but they use different ideologies and practices of governance to do so (Tan-Mullins, Mohan and Power, 2010). Claims that the "Beijing Consensus" constitutes a purposeful effort on the part of China to encourage autocratic development in opposition to "Western-style" development are also not supported by examination of the range of Chinese diplomatic activities. For example, China's participation in peacekeeping operations in Africa is supportive of transforming war-torn states into liberal democracies (Suzuki, 2009). Each of the pieces referred to here notes a tendency in Western policy analyses to "look for potential threats" in Chinese actions.

Perceptions of threat on the part of Westerners

Some of the representations of East Asian resource investment as threatening are from Western commentators. There are two main reasons Western commentators may find this investment problematic. One is a philosophical difference, with people believing the East Asian model, as they understand it, is not the "right" way to go about resource interdependence, as outlined in the above discussion about liberalism. Another is a sense of affront on the part of incumbents being challenged by a rising power. These reasons are often mixed together, with the latter being expressed in the discourses of the former. In other words, some allegations of threat from East Asian resource investment are best characterised as attempts to discredit a rising power by questioning the morality of its methods.

The trope of East Asian companies hungry for resources, disregarding the damage to the environment or local people and encouraging bad governance, has a long history in Western imaginaries. Judith Bennett (1987, pp. 206–10) has traced historical relations between Chinese settlers and European colonisers in Solomon Islands, showing that negative images of Chinese traders originated as early as the 1800s with Europeans resentful of the competition offered by Chinese businesses. Bennett found no evidence that the Chinese were any more exploitative of locals than were white colonials.

A sense of rivalry in response to the rise of Japan on the part of Westerners was visible from the late 1970s in a range of books such as Ezra Vogel's (1979) *Japan as Number One: Lessons for America*, Bill Emmot's (1989) *The Sun Also Sets: Why Japan Will Not Be Number One* and William Nester's (1990) *Japan's Growing Power over East Asia and the World Economy*. Some of this literature expresses rivalry mixed with admiration, while some of it attempts to discredit Japanese success as wrong in various ways. During and after the 1997 Asian crisis, some Westerners triumphantly decried Japanese "crony capitalism" (Milner, 2000). This sense of rivalry carried through into Western perceptions of Japanese resource industries, notably fisheries (Barclay, 2007; 2012).

On the other hand, a sense of resentment against newcomer non-Western powers is not the only influence on observations of East Asian resource investment. Western perceptions of Japanese investment, for example, have been coloured by negative associations from Japan's role in World War II. Western understandings of Chinese investment are tinged by the Cold War and China's history of opposition to liberal capitalism, and its continuing authoritarian style of government. Clearly many instances of East Asian resource investment have not been as socially or environmentally responsible as they should have been, just like Western resource investment. And self-interest plays a role in East Asian countries, securing supplies of resources, just as it does in Western countries' approaches to resource supplies. Furthermore, some of the denunciations of Chinese resource investment as mercantilist come, ironically, from the original East Asian mercantilist state, Japan. On this point we would argue that despite Japan's undoubted status as East Asian, it has also had an ambiguous status as Western, due to its early modernisation and industrialisation. In this sense Japan, as an incumbent power responding unfavourably to the rise of new power China, shares the perspective of the West.

The tendency on the part of many Western analysts to find a threat when the evidence is arguably insufficient to justify concern is significant. Insofar as policy is influenced by such ideas they have the potential to unnecessarily build tensions between Western powers and China. It would be much more helpful if Western commentators and policymakers resigned themselves to the end of Western domination and responded to East Asian investment in resources on its merits – in all its diversity – rather than continuing to build upon preconceived notions of threat.

Former colonies' perceptions of threat from Asian investment

Western commentators accustomed to viewing the rest of the world as subordinate, however, are not the only ones who have found East Asian resource investment to be threatening. Many former colonies of the Western powers also see East Asian resource investment as threatening. With the dismantling of formal militarily controlled colonies in the second half of the twentieth century, anti-colonial activists found that continued imperial domination went on through the influence of corporations from former colonising states in the economies of former colonies. This process, involving entwined processes of cultural assimilation, asymmetrical economic interdependence and socio-political influence, has been called neo-colonialism. The liberal pillars of global capitalism, trade liberalisation and associated political changes pushed by aid donors and international financial institutions such as the World Bank are some of the main conduits by which neo-colonialism is believed to occur. While the term was first applied to former European colonial powers, East Asian investment has also been described as neo-colonial.

In Africa, fears of Chinese interests using their powerful position in developing countries to unfairly pursue their own interests at the expense of the host countries are complicated by China's history of anti-colonial activity in Africa dating from the 1950s. Some African elites actively promote this history of cooperation as part of their technique for defending and building support for contemporary Chinese investment (Large, 2008). In the islands of the Pacific, however, China has no such history of anti-colonial cooperation, and consequently is more straightforwardly seen as potentially neo-colonial (Smith, 2012).

China's strong commitment to the principle of political non-intervention in some cases is thought to make it less colonial than Western countries. Former colonial powers have been willing to use their economic power over aid recipients to intervene in their domestic politics in the name of improving "governance". Former colonies are understandably sensitive about their sovereignty and thus appreciate this less domineering aspect of Chinese involvement (Moyo, 2009).

The means of ameliorating sensitivities about a threat from China thus differ depending on whether one is dealing with former colonies or Western powers. Western countries usually want China to adhere to liberal norms – placing trust in world trade to provide supplies of necessary resources and participating in regimes that support the international trade in resources – to make China less threatening. China's "charm offensive" deployed as a soft-power tool to allay fears of a "China threat" in the West has involved demonstrating its liberal credentials (Suzuki, 2009). Demonstrating liberal credentials, however, will not allay the fears of former colonies that China might be a latent neo-colonialist. Indeed, open participation in world markets and adherence to the norms of international regimes are keys to the operation of neo-colonialism, so being more liberal will in no way reduce the likelihood of China being neo-colonial. China needs quite a different charm offensive to allay fears of a China threat in former colonies. The history of Chinese South–South cooperation is a good background for this charm offensive but as China is increasingly viewed as a powerful and wealthy country it needs new ways to demonstrate that it will not dominate smaller countries. One difficulty with this is that the two charm offensives tend to counteract each other. As seen above in the debates over China's political non-interference in resource-producing countries, efforts not to dominate politically in developing countries may result in increased perceptions of threat on the part of Western powers. That is, Western powers want China to influence the politics of resource-producing countries in alignment with the good governance agenda, so it will be difficult for China to satisfy those demands while simultaneously demonstrating to resource-producing countries that it is not.

Interestingly, Korea has not been considered a threat in the way Japan was in the 1980s and China is now. This difference has not come about because Korean resource investment has been any more ethical or in tune with local norms than that of the other two. For example, in 2008 Daewoo Logistics did a deal with the Madagascar Government for a 99-year lease for 3.2 million hectares of land – half the arable land in the whole country – for the purpose of growing corn and palm oil. Existing public dissatisfaction with the Madagascar Government of the day was fuelled by outrage over this deal, resulting in a coup d'état. The new government cancelled the deal with Daewoo (Oviedo, 2011). For explanations of why a "Korea threat" thesis has not gained traction, we look rather to broader political and historical factors: for instance, the fact that it is a middle power and thus not seen as potentially dominating other countries in the same way as the larger powers Japan and China. Furthermore, Korea does not have Japan's history of imperial expansion; nor, until recently, has the Korean state been obviously connected to the overseas resource investments of Korean *chaebol*. The actions of Korean corporations have therefore not been seen as the actions of "Korea" in the same way as the actions of state-owned Chinese oil companies may be seen as the actions of "China", for example.

Conclusion

In this Special Issue the substantive contributions consider a series of questions that are foreshadowed in this introductory paper. First, some of the papers examine the question of how East Asian governments and companies actually behave in resource-producing developing countries. James Reilly traces the history of patterns of development assistance in relation to resource investment, which were first employed by Japan, and are now employed by China. He studies the international political consequences of competitive aid and resource investment between China and Japan in Myanmar. The paper by Philippa Brant takes scholarship on Chinese aid and resource investment in Africa and looks at whether Chinese aid and resource investment in Pacific Island countries is following similar trajectories. Specifically, how directly connected are Chinese foreign policy and resource investment? Is it accurate to speak of private sector investment being part of a state-led form of resource diplomacy? Graeme Smith then looks at Chinese involvement in mining in Papua New Guinea. How do the labour strategies vary from those in Western mining investments, and what are the consequences for Papua New Guinean society and the employees involved?

A second major set of questions addressed by the papers in this issue is how East Asian governments treat the role of markets and international regimes in resource interdependence. Ben Saul's paper asks how China measures up in terms of abiding by international law. Kate Barclay and Charlotte Epstein explore the Japanese Government's balancing act between world markets and government support for domestic producers in policies for food security, and Japan's positioning in international organisations to promote these policies. Ming Hwa Ting and John Seaman look at how the unfolding of markets in rare earth minerals has led to the situation where China has a virtual monopoly on supply. The Chinese Government has at times restricted exports for political reasons, and the fallout from this may affect territorial politics as Japan and South Korea, both heavily dependent on rare earths for their high technology industries, look to secure claims over seabed areas with potential for rare earth mining. Given that East Asian governments and companies are not the only drivers in the production of and trade in resources, what are some of the other influential factors? Luke Nottage's paper shows how an Australian policy on protection of international investor interests may affect the general environment of cross-border investment in resources in the Asia Pacific region.

The international politics of resources in East Asia has become increasingly important in recent years as the world economy has come to rely heavily on development in Asia to cover for the stagnation occurring in Europe and the United States. Resources are the raw ingredients of economic growth in those economies, at least as they are currently structured, so this topic will be of major significance for the foreseeable future.

Acknowledgments

Many thanks to the University of Technology Sydney China Research Centre and the University of Sydney China Studies Centre for sponsoring the workshop at which this collection of papers was first presented. Thanks are due to all workshop participants for their comments on the papers, with special mention to Steve Fitzgerald, Peter Drysdale and Michael Wesley for helping to develop conceptual frameworks for the workshop. Thanks also to two anonymous reviewers of this paper, whose comments helped us refine our arguments.

Notes

1. "The West" and "Asia" are highly contestable and imprecise terms. We use "the West" here as shorthand for the countries of Europe, North America, Australia and New Zealand. For the purposes of this paper we do not include the former colonies of Latin America in "the West", as they arguably have a different standpoint because of their experiences of having been colonised. "Asia" is used here in a broad sense to include countries east of the Bosphorus River, while "East Asia" means the countries on which this Special Issue focuses – China, Japan and South Korea.
2. In a subsequent publication, *Brand China*, Ramo is a good deal less upbeat, concluding that China is not trusted abroad, its brand is weak, and there is a need for self-invention that is not a "whitewash" (Ramo, 2007).
3. 'China's Foreign Policies for Pursuing Peaceful Development'. Available at http://www.gov.cn/english/official/2011-09/06/content_1941354_4.htm, accessed 6 November 2011.
4. In PNG, a Chinese state-owned enterprise is a new player in the mining scene (see Smith, this volume). In both Solomon Islands and PNG, ethnic Chinese business people have been the main buyers for seafood exports (mainly dried sea cucumber, shark fin and shells) for more than a century, and in recent decades Taiwanese and "new" Chinese companies have been key players in fishing for tuna and shark in these countries.

References

Barclay, Kate (2007) Fishing. Western, Japanese and Islander perceptions of ecology and modernization in the Pacific. *Asia Pacific Journal: Japan Focus*. Available at http://www.japanfocus.org/-Kate-Barclay/2508, accessed 20 August 2010.

Barclay, Kate (2012) Development and negative constructions of ethnic identity: Responses to Asian fisheries investment in the Pacific. *The Contemporary Pacific* 24(1), pp. 33–63.

Bell, C. (2007) *The end of the Vasco da Gama era: The next landscape of world politics* (Sydney: Lowy Institute for International Policy).

Bennett, J. (1987) *Wealth of the Solomons: A history of a Pacific archipelago, 1800–1978*. Pacific Islands Monograph Series No. 3 (Honolulu: University of Hawai'i Press).

Bräutigam, Deborah A. (2010) *The dragon's gift: The real story of China in Africa* (Oxford: Oxford University Press).

Breslin, Shaun (2011) The "China model" and the global crisis: From Friedrich List to a Chinese mode of governance? *International Affairs* 87(6), pp. 1323–43.

Buijs, B. and H. Sievers (2011) *Critical thinking about critical minerals*. CIEP–BGR Briefing Paper. Available at http://www.clingendael.nl/publications/2011/20111111_ciep_bgr_briefingpaper_buijs_sievers_critical_ thinking_about_critical_minerals.pdf, accessed 11 December 2011.

Brown, Lester R. (1995) *Who will feed China? Wake-up call for a small planet* (Washington, DC: Worldwatch Institute).

Buruma, Ian (2008) China's dark triumph: The success of its economy poses a serious challenge to liberal democracy. *Los Angeles Times*, 13 January.

Callick, Rowan (2007) China's neo-colonial slavery in PNG. *The Australian*, 12 February. Available at http://www.theaustralian.com.au/business/chinas-neo-colonial-slavery-in-png/story-e6frg8zx-1111112977807, accessed 28 February 2013.

Dauvergne, P. (1997) *Shadows in the forest: Japan and the politics of timber in Southeast Asia* (Cambridge, MA: MIT Press).

Deng, G. and S. Kennedy (2010) Big business and industry association lobbying in China: The paradox of contrasting styles. *The China Journal* 63, pp. 101–25.

Emmot, Bill (1989) *The sun also sets: Why Japan will not be number one* (London: Simon & Schuster).

Grilli, E.R. (1996) *The European Community and the developing countries* (Cambridge, UK: Cambridge University Press).

Haglund, Dan (2009) In it for the long term? Governance and learning among Chinese investors in Zambia's copper sector. *The China Quarterly* 199, pp. 627–46.

Hanson, Fergus (2009) *China: Stumbling through the Pacific* (Sydney: Lowy Institute for International Policy).

Ikenberry, John G. (2010) The liberal international order and its discontents. *Millennium: Journal of International Studies* 38(3), pp. 509–21.

Jiang, Julie and Jonathan Sinton (2011) *Overseas investments by Chinese national oil companies*. IEA. Available at http://www.iea.org/publications/freepublications/publication/name,3947,en.html, accessed 28 February 2013.

Johnson, Chalmers (1982) *MITI and the Japanese miracle* (Stanford, CA: Stanford University Press).

Kennedy, Scott (2010a) The myth of the Beijing consensus. *Journal of Contemporary China* 19(6), pp. 461–77.

Kennedy, Scott (2010b) The Mandarin learning curve: How China is reshaping global governance, in *China Social Science Workshop* (Stanford, CA: Stanford University)).

Kim, Joongi (2008) A forensic study of Daewoo's corporate governance: Does responsibility for the meltdown solely lie with the chaebol and Korea? *Northwestern Journal of International Law & Business* 28(2), pp. 273–340.

Large, Daniel (2008) Beyond "dragon in the bush": The study of China–Africa relations. *African Affairs* 107, pp. 45–61.

Lee, Chin Kwan (2009) Raw encounters: Chinese managers, African workers and the politics of casualization in Africa's Chinese enclaves. *The China Quarterly* 199, pp. 647–66.

Lee, John (2012) China's geostrategic search for oil. *Washington Quarterly* 35(3), pp. 75–92.

Lister, M. (1988) *The European Community and the developing world: The role of the Lome Convention* (Aldershot: Avebury).

Mbaye, Sanou (2011) Africa will not put up with a colonialist China. *The Guardian*. Available at http://www.guardian.co.uk/commentisfree/2011/feb/07/china-exploitation-africa-industry/print, accessed 28 February 2013.

McCarthy, Joe (2013) Crude oil-mercantilism? Chinese oil-engagement in Kazakhstan. *Pacific Affairs* 85(3), forthcoming.

McKibbin, W., M. Cook, M. Fullilove, M. Wesley, A. Bubalo, R. Medcalf, F. Hanson, J. Hayward-Jones and S. Roggeveen (2010) How the roar of China's tiger will be heard across the world. *The Australian*, 13 February, Inquirer, p. 2.

Milner, A. (2000) What happened to "Asian values"? in G. Segal and D.S.G. Goodman (eds), *Towards recovery in Pacific Asia*, pp. 56–68 (London: Routledge).

Ministry of Foreign Affairs and Trade, Republic of Korea (n.d.) *Korea's ODA at a glance*. Available at http://www.odakorea.go.kr/eng/include/glance.php, accessed 7 December 2011.

Moyo, Dambisa (2009) *Dead aid: Why aid is not working and how there is a better way for Africa* (London: Allen Lane).

Nester, William (1990) *Japan's growing power over East Asia and the world economy* (New York: St Martin's Press).

Nottage, Luke (2013) Investor-state arbitration policy and practice after Philip Morris Asia v Australia, in Leon Trakman and Nicola Ranieri (eds), *Regionalism in international investment law*, pp. 452–74 (Oxford: Oxford University Press).

Oviedo, Sheila (2011) Avoiding the land grab: Responsible farming investment in developing nations (Amsterdam: Sustainanalytics). Available at http://www.sustainalytics.com, accessed 8 December 2011.

Pinstrup-Andersen, Per (2009) Food security: Definition and measurement. *Food Security* 1, pp. 5–7.

Ramo, Joshua Cooper (2004) *The Beijing consensus: Notes on the new physics of Chinese power* (London: The Foreign Policy Centre). Available at http://www.fpc.org.uk/fsblob/244.pdf, accessed 6 November 2011.

Ramo, Joshua Cooper (2007) *Brand China* (London: The Foreign Policy Centre). Available at http://fpc.org.uk/fsblob/827.pdf, accessed 28 February 2013.

Ravenhill, John (1985) *Collective clientelism: The Lome Conventions and North–South relations* (New York: Columbia University Press).

Reilly, James (2012) A Northeast Asian model of ODA? Comparing Chinese, Japanese and Korean official development assistance, in Joern Dorsch and Christopher Dent (eds), *The Asia-Pacific, regionalism and the global system* (London: Edward Elgar).

Renton, Alex (2005) One in ten fish is eaten in Japan. So why don't they know there's a shortage? *Observer Food Monthly*, 10 April. Available at http://www.guardian.co.uk, accessed 19 August 2010.

Rosen, Daniel H. and Thilo Hanemann (2009) *China's changing outbound foreign direct investment profile: Drivers and policy implications* (Washington, DC: Peterson Institute for International Economics).

Sautman, B. and H. Yan (2008) The forest for the trees: Trade, investment and the China-in-Africa discourse. *Pacific Affairs* 81(1), pp. 9–29.

Smith, Graeme (2012) Chinese reactions to anti-Asian riots in the Pacific. *Journal of Pacific History* 47(1), pp. 93–109.

Suzuki, Shogo (2009) Chinese soft power, insecurity studies, myopia and fantasy. *Third World Quarterly* 30 (4), pp. 779–93.

Tan-Mullins, May, Giles Mohan and Marcus Power (2010) Redefining "aid" in the China–Africa context. *Development and Change* 41(5), pp. 857–81.

Tarte, S. (1998) *Japan's aid diplomacy and the Pacific Islands* (Canberra: Asia Pacific Press).

Tsing, Anna Lowenhaupt (2005) *Friction: An ethnography of global connection* (Princeton and Oxford: Princeton University Press).

UNDP (2000) *World energy assessment: Energy and the challenge of sustainability* (New York: United Nations Development Programme [UNDP]; United Nations Department of Economic and Social Affairs; World Energy Council).

Vogel, Ezra (1979) *Japan as number one: Lessons for America* (Cambridge, MA: Harvard University Press).

Wesley-Smith, Terence and Edgar A. Porter (2010) *China in Oceania: Reshaping the Pacific?* (New York: Berghahn Books).

Williamson, John (1990) What Washington means by policy reform, in John Williamson (ed.), *Latin American adjustment: How much has happened?* (Washington, DC: Institute for International Economics).

Windybank, Susan (2005) The China syndrome. *Policy* 21(2), pp. 1–6. Available at http://www.cis.org.au/images/stories/policy-magazine/2005-winter/2005-21-2-susan-windybank.pdf, accessed 28 February 2013.

Zhu, Feng (2011) *The limits of power: Why China is a "bad neighbour"*. Available at http://www.first-post.com/world/the-limits-of-power-why-china-is-a-bad-neighbour-120299.html, accessed 6 November 2011.

China and Japan in Myanmar: Aid, Natural Resources and Influence

JAMES REILLY

The University of Sydney

Abstract: *This article highlights the parallels between Japan's leveraging of foreign aid to secure access to strategic natural resources in the post-World War II period, and China's similar use of foreign aid since the early 1990s. It then shows how both countries have applied similar strategies in their aid programs in Myanmar, and points out how dramatic shifts in Myanmar's domestic politics have shaped their opportunities for economic engagement. China seized upon Myanmar's political transition in 1988 to expand its aid program, while Japan has responded to Myanmar's reform measures since 2010 by resuming its aid efforts. The article concludes by considering the potential for China–Japan cooperation through their aid programs in Myanmar.*

China has been widely criticised for using foreign aid to secure access to strategic natural resources. Less frequently noted are the parallels with Japan's earlier efforts to leverage its foreign aid program in pursuit of energy resources. In the early 1970s, Japan's rapidly growing industrial sector required growing levels of imported oil and other energy sources. The oil shocks of 1973 and 1979 were a rude awakening for a Japanese economy that had become dependent upon steady access to cheap oil under the American imperium. Flush with US dollar reserves, Japan successfully deployed its burgeoning foreign aid program to secure access to oil shipments from the Middle East. Japan's official development assistance (ODA) to China, which began in 1979, also obtained access to Chinese coal and oil in exchange for building the infrastructure necessary to extract and export these resources back to Japan. Only in the early 1990s did Japan's pragmatic, neo-mercantilist approach to foreign aid slowly begin to shift in response to international and domestic pressures.

Around the same time, China's own aid program began to re-emerge. China had been a substantial provider of development assistance to the developing world throughout the Maoist era. The onset of reform in 1978 led to a drawdown in China's aid levels, as Beijing shifted toward a more pragmatic focus upon developing its own economy. By the mid-1990s, China faced a situation similar to Japan in the 1970s: a booming industrial economy that was flush with cash due to expanding exports, but increasingly dependent upon imported oil and other strategic resources. Like their Japanese counterparts two decades earlier, Chinese leaders began to leverage foreign aid as part of a broader strategy to secure access to strategic natural resources in developing countries. While China's aid program varies from Japanese aid in many respects, this parallel is striking.

China and Japan are hardly unique in using foreign aid to advance their strategic interests. Foreign aid has generally been used to serve donor countries' strategic, economic and diplomatic objectives, as well as domestic interests (Lancaster, 2006, pp. 5–8). Directing public and private investment toward strategic natural resources has historically been a valuable strategy for states "defending the national interest" (Krasner, 1978, p. 13). The China–Japan comparison is, however, particularly interesting for its timing – Japan began to shift away from explicitly linking its development assistance to resource extraction just as China adopted this approach. One of the most important examples of this passing of the guard has been in Myanmar (Burma). Propelled by its postwar reparations, Japan had been deeply engaged in Myanmar for decades – until the events of 1988 that brought the current ruling junta to power led to a rapid drop in Japanese investment and aid. China quickly moved in to pick up where Tokyo left off. Beijing bundled aid and investment to enhance Chinese access to key natural resources. Beginning in 2010, another round of dramatic political changes in Myanmar provided an opening for Japan to resume its economic engagement with Myanmar while China's role in the country was coming in for greater criticism.

This article begins by reviewing Japan's aid program, focusing upon its pursuit of natural resources through aid provision in the 1970s. The second section examines China's aid program following its reforms of the mid-1990s and highlighting China's pursuit of natural resources. It then turns to the case of Myanmar, tracing the history of Japanese and Chinese aid programs in the country before concluding by weighing the potential for cooperation in China and Japan's aid programs in Myanmar.

Japan as Number One

Japan's foreign aid program originated out of the provision of reparations to Asian countries following the San Francisco Treaty of 1951, and subsequent bilateral agreements reached with Burma (1955), the Philippines (1956) and Indonesia (1958). The close linkage between foreign aid and reparations had two important long-term effects upon Japan's aid program. First, Japan was extremely reluctant to impose political conditions upon its aid recipients. This was evident particularly in Japanese aid to China, but it also characterised Japanese aid to Burma, Indonesia, South Korea and elsewhere. As late as 1990, Japan's Ministry of Foreign Affairs (MFA) explicitly equated such conditions with engaging in undue interference in the internal affairs of recipient governments (Takamine, 2006, p. 73). Secondly, Japanese aid has traditionally been oriented toward its Asian neighbours. Up until 1972, some 98 per cent of all Japanese aid was directed to the Asia Pacific. By 1990, Asia still received 59 per cent of all Japanese aid (Koppel and Orr, 1993, p. 5).

Directing its aid to Asia made good economic sense for Japan as it sought to restart its postwar economy and expand its exports to the region. Japan's reparations were thus tied to procurement of Japanese products. The formal launch of its "economic cooperation" program in 1961 coincided with Prime Minister Ikeda Hayato's plan to double Japan's national income in a decade. Throughout much of this period, Japan's aid was characterised by several distinct elements. First, Japan relied heavily upon the provision of concessional loans, which began in 1957. This was largely because ODA funds drew from the Fiscal Investment and Loan Program (FILP), which was largely funded by the postal savings of Japanese citizens, and so had to be repaid in full. The loans went mainly to Asia, and besides meeting certain needs in recipient countries, served to help establish Japanese industry in the area (Söderberg, 2002, p. 3). Japanese aid officials defended their heavy use of concessional loans on the basis that they enforced discipline among recipients, and so encouraged their adaptation to transnational market forces (Hook and Zhang, 1998, p. 1052). By the 1980s, Japan's habit of combining private investment, market-rate loans and concessional loans all under the category of "economic cooperation" was "causing confusion in Western capitals, especially when Japan makes major aid pledges at annual Group of Seven summits" (Koppel and Orr, 1993, p. 3).

Japan's aid was dominated by support for large-scale infrastructure: bridges, dams, highways, airports and port facilities. This was grounded in an assumption that the major contribution of external ODA was to provide such capital-intensive investments, laying a foundation for private investment and entrepreneurism to flourish. Japanese firms endorsed the concentration of aid programs on building economic infrastructure in recipient states, and in turn greatly benefited from this approach (Hook and Zhang, 1998, p. 1055). As Margee M. Ensign (1992, p. 175) explains, Japan's aid program reflected its "unique brand of capitalism" in which "the state is more active and interventionist: business and government work closely together to nurture trade and industrial policies". Japan also provided aid within the context of its tradition of "gift-giving diplomacy" (*omiyage gaikō*), aimed at bolstering diplomatic ties (Katada, 2010, p. 56).

Japanese development aid was "explicitly regarded as a legitimate arm of national policy" (Rix, 1980, p. 268). The strategic use of aid was particularly evident in Japan's resource diplomacy of the 1970s, which provided aid and investment to countries possessing resources essential to Japanese development (Okita, 1974, p. 720). Japan's dependence upon imported natural resources was acute. By 1972, Japan imported 58 per cent of its timber consumption, 92 per cent of its iron ore, 59 per cent of its coal, 100 per cent of its bauxite, 84 per cent of its copper, and 99.7 per cent of its oil (Potter, 1992, p. 68). The "soybean shock" of 1973, followed by the oil embargo resulting from the 1973 Yom Kippur War, heightened Japan's sense of vulnerability. Tokyo responded by rapidly increasing its aid to the OPEC states in the Middle East, successfully encouraging them to lift the oil embargo for Japan (Nester and Ampiah, 1989, p. 78). Funding for major resource extraction infrastructure projects in oil producing countries, such as the Iran–Japan Bandarkhomeini petrochemical project, resulted in Asia's share of Japan's ODA falling from nearly 100 per cent to two-thirds over the next few years (Ott, 1985, p. 230). Japan's Export–Import Bank – the primary source of finance for yen loans – also increased its credit for resource development projects in other supplier countries, such as Japanese public–private collaboration in developing the oil industry in Indonesia. Japanese investments into infrastructure were generally repaid in kind with oil exports. A similar model of finance and barter trade was applied to projects of nickel exploitation in Indonesia and the Philippines, and the Asaha aluminium refinery complex in Indonesia (Potter, 1992, Chapters 4 and 6).

Aiding China

Japan's aid to China followed a similar approach. Following the normalisation of Sino-Japanese relations in 1972, the two governments signed four economic agreements: on trade (January 1974), civil aviation (April 1974), maritime transport (November 1974) and fishing (August 1975). In the Long-Term Trade Agreement, reached on 16 February 1978, China pledged to import US$10 billion worth of Japanese technology and capital equipment over the next eight years, to be paid for by exporting Chinese oil and coal to Japan. Following the signing of the Treaty of Peace and Friendship on 12 August 1978, Japan issued its first round of yen loans to China, pledging 330 billion yen (US$1.4 billion) for 1979–83 (Wan, 2006, p. 21). By the end of 1978, Japan had signed 74 contracts with China to finance turnkey projects – all to be repaid in oil. Japan's yen loans went first and foremost to the construction of railways and ports to ensure the export of coal and oil to Japan. They subsequently funded hydroelectric and thermal power plants, urban water supply, telecommunications, highways and fertiliser plants – sectors that formed the backbone of China's modernisation (Bräutigam, 2009, p. 51). Key projects included a major petrochemical plant at Daqing and an iron mill at Baoshan. In a symbol of friendship, Japan also provided grant aid to fund the establishment of a 1,000-bed hospital in Beijing – the Sino-Japanese Friendship Hospital. Japan's second aid package (1984–89; 540 billion yen) was also primarily financing for infrastructure, particularly in support of extraction and transportation of natural resources (Söderberg, 2002, p. 8). By 1982, China had become the top recipient of Japan's bilateral ODA, while Japan emerged as the leading bilateral donor to China (Brooks and Orr, 1985, p. 335).

Not wishing to miss out, Western nations quickly followed Japan's example in their dealings with China, establishing compensatory trade (*buchang maoyi*) in which the Chinese company first imported foreign equipment and machinery while paying later with raw materials. As a sweetener, they too offered low-interest loans and aid projects. As Deborah Bräutigam (2009, p. 51) explains:

> China saw all of these tactics as *beneficial* for China's development. Japan and the West could use their modern technologies to exploit natural resources that Chinese technology could not yet unlock. China could pay for this investment later, with the resources that were uncovered. The subsidies and aid used by the West and Japan to wrap their naked hunger for China's markets meant that China was getting a discount on finance the country needed for its modernization.

Yet changes were afoot. As its overall aid levels continued to rise, Japan faced domestic and international pressure to revamp its approach to ODA. In June 1992, Prime Minister Miyazawa Kiichi approved a new ODA charter which identified four priorities for Japan's aid program: (1) environmental conservation; (2) the promotion of democracy and human rights; (3) restraints on military expenditures, the development of weapons of mass destruction, and arms transfers; and (4) the introduction of a market-oriented economy. The Japanese Government further pledged that due consideration would be paid in particular to least-developed countries (Government of Japan, 1992). While the extent and speed with which Japan enacted these pledges was widely questioned (Arase, 1995; Ensign, 1992; Hook and Zhang, 1998), the new approach soon began to affect Japanese aid to China.

By the mid-1990s, public opinion in Japan had shifted decisively against economic assistance to China. Japan's economic assistance appeared to have purchased scant diplomatic goodwill and instead seemed to be supporting China's military expansion. In 2005, after several years of sharp decline in aid levels, Japanese diplomats signalled their intent to end all yen loans to China by 2008. China responded surprisingly gracefully, with a Foreign Ministry spokesperson expressing China's hope that the ODA program would simply "start well and end well" (Wan, 2006, p. 285). Just as China was graduating from Japan's aid program, it was busy revamping and expanding its own overseas assistance.

China's Aid Program

China is one of the world's most experienced providers of foreign assistance, with an aid program dating back to 1950. Finding itself at loggerheads with both the Soviet Union and the United States by the 1960s, China rapidly expanded its aid program in support of new nations "fighting against colonialism and hegemony" (*fanzhi fanba*) (Shang, 2010, p. 57). By 1973, Beijing was providing aid to seven countries in Asia, six in the Middle East, three in Latin America, and 29 in Africa (Bräutigam, 2009, p. 41). The costs skyrocketed. From 1955 through to 1979, China's ODA expenditures averaged 0.87 per cent of GDP and 2.98 per cent of total government expenditures. By 1973, ODA spending reached 2.052 per cent of China's GDP, taking up an astonishing 6.9 per cent of total government expenditures – more than 25 per cent larger than the educational budget in that year. China had aid projects in more countries in Africa than even the United States (Bräutigam, 2009, p. 41).

Clearly, the scale of the program had become unsustainable. A 1975 State Council conference signalled a drawdown in assistance by urging that aid recipients "rely on themselves" in developing their economies (Yang Fangyuan, 2009, p. 231). Efforts were soon launched to bring the aid program under control, and aid levels began to fall. At the same time, Chinese leaders began to seek mutual benefits through foreign aid. In July 1979, Deng Xiaoping told a Politburo working group: "we must ensure that both the donor and the recipient country can receive benefits" (Bin, 2008, p. 36). The following year, a State Council report declared that the aid program would: "maintain proletarian internationalism, maintain the eight principles of foreign aid, [ensure] mutual benefit, promote the economic development of friendly nations, and speed up the realization of the four modernizations in China" (cited in Bin, 2008, p. 36). Two years later, Premier Zhao Ziyang announced four guiding principles for China's economic relations with developing countries: equality and mutual benefit, a stress on practical results, diversity in form, and common progress (Bräutigam, 2009, p. 50).

In the mid-1990s, China instituted a series of reforms in its aid program. As Kobayashi (2008, pp. 14–15) explains, three key pillars of the aid program emerged. First, the Export–Import Bank, established in 1994 as one of China's three policy banks, became the major financer for China's burgeoning concessional loan program. Secondly, foreign aid became more closely integrated with China's external economic strategy. The State Council's economic and trade strategy of 1995 explicitly linked foreign aid with economic cooperation and trade. Loans were expected to finance joint ventures in manufacturing and agriculture, with aid provided to "promote the export of our medium and small equipment, processing machinery, relevant technology, and labor service" (cited in Kobayashi, 2008, p. 41). Finally, diplomatic heavyweights got behind the aid program. After Zhou Enlai's tours of the early 1960s, there was only one high-profile Chinese tour of Africa: Zhao Ziyang's visit to 10 African countries in 1982–83. This hiatus ended abruptly in 1995, as three different Chinese Vice-Premiers, including future Premier Zhu Rongji, visited a total of 18 African countries. Since then, leaders' visits have been a key part of China's aid program in the developing world. President Hu Jintao, for instance, has been to Africa six times. All three attributes were soon deployed to serve China's energy needs.

By the early 1990s, China was facing an energy shortage. China became a net importer of petroleum in 1993 and of crude oil in 1996. China overtook Japan as the second largest oil consumer next to the US in 2003, and became the third largest oil importer after the US and Japan in 2004 (Zweig and Bi, 2005, p. 30). From 2000 to 2005, China's energy consumption rose by 60 per cent, accounting for almost half of the growth in world energy consumption. Although China is able to meet more than 90 per cent of its primary energy requirement with domestic supplies, it still imports almost half of the oil it consumes. Zheng Bijian, a senior advisor to Chinese President Hu Jintao, described the shortage of resources as the first of three fundamental challenges for China's peaceful rise (Zhao, 2008, p. 211).

China soon began to leverage its burgeoning capital resources to secure access to strategic natural resources, with its first subsidised loan to Sudan in 1996 to finance oil exploration through a joint venture with China National Offshore Oil Corporation. Two years later, the CCP Central Committee urged an expansion of China's foreign direct investment "particularly in Africa, Central Asia, Middle East and Central and Latin America". From 2000 through 2005, China's outward FDI to Africa and Latin America rose by 37 per cent and 45 per cent respectively, with large increases in mining – particularly the oil sector (Kobayashi, 2008, p. 40). The Tenth Five Year Plan (2001–05) described the acquisition of natural resources as essential for China's development, and issued the first call for Chinese firms to "go global" (*zou ququ*). In addition to a stronger emphasis upon "ensuring economic security" through access to strategic natural resources, the Eleventh Five Year Plan (2006–10) pledged an increase in aid as well as economic and technical cooperation with other developing countries (Kobayashi, 2008, p. 40).

In the widely cited Angola model, applied in sub-Saharan Africa and Central Asia, the Chinese Government first reaches an agreement on provision of a subsidised loan to a resource-rich country in order to construct infrastructure essential for resource development. The Ministry of Commerce then holds a competitive bidding process to select which Chinese construction company will implement the project. Regulations specify that Chinese companies complete the infrastructure projects funded by these concessional loans. Funds are dispersed directly to Chinese firms, which then implement the aid project. This practice helps to mitigate local corruption and helps to ensure that projects are relatively efficient and cost-effective (Kobayashi, 2008, p. 22). In exchange, the borrowing government provides a Chinese state-owned corporation the access rights to extract the minerals either through equity shares in a national oil company or through acquiring licences for production. The recipient country puts up its resources – oil, gold, gas – as security for the subsidised loan. This model has been used to support the "Go Out" strategy in which Chinese state companies are encouraged to invest abroad, enter foreign markets, and establish manufacturing or mining facilities overseas (Aquino and Jensen-Joson, 2009, p. 41). Similar to Japan's earlier aid program in China, Chinese subsidised loans are often part of a larger investment package aimed at securing access to key strategic resources in developing countries (Zhang, 2010, p. 25).

To be clear, the bulk of China's aid program is not directed to resource extraction. Most of its resource-related loans are on a commercial basis, and even the subsidised loans would likely not qualify as ODA under OECD standards (Bräutigam, 2010, p. 5). Nor, for the most part, are these arrangements presented as development aid by Beijing. China's 2006 White Paper on foreign aid explains that China "makes active cooperation with the overseas expansion strategy of firms for going global and builds a platform for Chinese firms to open up markets, develop resources, and export high-tech products abroad" (Kobayashi, 2008, p. 41). While acknowledging that China is engaging in "resource diplomacy", the Director General of the Department of Foreign Economic Cooperation insists: "Chinese external resource cooperation has been consistently conducted under the win-win principle". He argues that China is enabling resource-endowed countries to obtain wealth, transferring resource exploration and development technologies to these countries, enhancing their employment and revenues by processing resource-based products in these countries, and helping achieve bilateral trade balances (*South China Morning Post*, 20 August 2007, A3). China's Export–Import Bank describes foreign aid as a "vanguard" supporting Chinese exports and investments while contributing to sectors such as transportation, telecommunications and energy, "thus improving the investment environment in developing countries" (Export–Import Bank, 2006, p. 21).

Chinese scholars describe this approach as practical and beneficial for both donor and recipient. "China's new approach to aid reduces the fiscal burden on our country, and supports our own economic development," explains Sun Luxi (2007, p. 9). "Strengthening trade, education, and other co-operative programmes through [our] foreign assistance enables our country and our enterprises to receive greater economic benefits." Zhai Dongsheng (2010, p. 92) adds, "Our national perspective is one of equality … the [aid] relationship is one of equality and of practical mutual benefit".

In sum, like Japan in the mid-1970s, China's burgeoning export-oriented industrial economy generated both a growing demand for energy imports and sufficient capital to purchase these resources. Like Japan, China's mercantilist approach is grounded in scepticism that these resources can be secured solely through international markets, and instead has relied upon strategic direction of government investment and aid to resource-rich states in order to diversify its sources of energy supplies. Similar to fears about Japan's rapid rise in the 1980s, China's pursuit of natural resources has given rise to deepening anxiety in the West. As one US scholar warned:

> The results of China's energy diplomacy are being watched with growing unease, especially in Asia but in other parts of the world as well ... There is a danger that China's neo-mercantilist strategy to bolster energy security by gaining direct control both of oil and gas fields and supply routes could result in escalating tensions in an already volatile region that lacks regional institutions for conflict resolution and is in the midst of a difficult transition process, which is due in fact to the rise of China. Competition for energy is exacerbating existing rivalries between China and a number of its neighbours (Kreft, 2006, n.p.).

One such potential rivalry yet to receive much international attention is Japanese and Chinese involvement in Myanmar. The case of Myanmar encapsulates both the similarities and differences in China and Japan's aid programs while highlighting the importance of domestic developments in shaping their involvement in Myanmar through aid provision.

Japan's Rise and Fall in Burma/Myanmar

In the 1950s, Japan emerged as a key player in Burma, leveraging its aid program to secure political influence and economic access. While Burma was the first recipient of Japanese postwar reparations, Prime Minister Yoshida Shigeru was determined that the reparations would benefit Japan. Yoshida later explained:

> Since the Burmese did not like the word investments, we used the word reparations as they wished, but for us it was investments. Through our investments Burma would develop and it would become a Japanese market, and so our investment would return (Edström, 2009, p. 19).

In November 1954, Japan and Burma reached an agreement on reparations of US$200 million to be paid out over 10 years, as well as US$5 million annually in Japanese goods and services for joint projects (Edström, 2009, p. 18). Economic relations between the two countries came to comprise trade, reparations and economic cooperation, the latter of which included investments and technical assistance. While the Burmese Government was desperate for the financial support, the Japanese side ensured that Japanese firms benefited from the arrangements. For instance, reparations funded the Baluchaung Dam project, conceived by the head of Nippon Koei Corp., Kubota Yutaka, who then persuaded Prime Minister Yoshida to support it. It was never requested by the Burmese Government and resulted in massive exports for Japanese industry (Edström, 2009, p. 20).

Japan's support became particularly important following the 1962 coup staged by General Ne Win and the Burmese armed forces, the Tatmadaw. Under their policies of the "Burmese Way to Socialism", encompassing militarism, nationalism, Buddhism and nationalisation, Burma's international isolation deepened. Throughout Ne Win's 16 years in power (1962–88), Burma was one of Japan's largest aid recipients, receiving a total of US$2.2 billion. By 1987, Japan's ODA constituted 20 per cent of Burma's national budget, making up 71.5 per cent of total foreign aid Burma received (Oishi and Furuoka, 2003, p. 899).

Following the brutal suppression of popular protests in 1988, Prime Minister Ne Win resigned, succeeded by a military junta. On 18 June 1989, the State Law and Order Restoration Council (SLORC) officially changed the country's name to the Union of Myanmar. Following these events, the Japanese Government joined international condemnation in suspending its aid program to Myanmar. Once the SLORC announced that general elections would be held in 1990, the Japanese Government in February 1989 recognised the military regime as the legitimate government and decided to resume aid to the country – pledging debt relief grants and small-scale humanitarian aid (Edström, 2009, p. 30). Yet as domestic and international pressure mounted for Japan to support international sanctions on Myanmar, and Japan's 1992 ODA Charter began to take effect, Japanese aid to Myanmar declined, particularly the subsidised loans. Under the Charter, Japan's Ministry of Finance is prohibited from providing new ODA loans if the recipient is in arrears on existing loans. In order to avoid Myanmar going into arrears, Japan provided some 68 billion yen (US$600 million) in debt relief to Myanmar from 1991 to 2003, representing 75 per cent of total Japanese ODA to Myanmar during this period. Despite this effort, Myanmar went into arrears in 1995, limiting Japan's capacity to provide further assistance (Streeford, 2007, p. 163). Japan's aid slowed substantially – limited primarily to small-scale grant aid provided as humanitarian assistance. Investment also dropped rapidly. From 1989 through 2011, Japanese investment in Myanmar totalled just US$212 million (Foster, 2012). As Japan's economic involvement waned, China rapidly emerged as Myanmar's most important Northeast Asian partner.

China's Deepening Engagement in Myanmar

Following the events of 1988–89, China was the first country to recognise the new regime. China soon emerged as one of Myanmar's most important sources of investment, trade and aid. While Myanmar's exports to China increased just 2.2 times, from US$126 million in 1989 to US$274 million in 2005, its imports from China grew nearly five-fold during the same period, from US$188 million to US$935 million, yielding a chronic trade deficit with China (Aung Myoe, 2007, p. 66). By 2009 bilateral trade totalled US$2.9 billion. By January 2010, China's investment in Myanmar amounted to US$1.8 billion, accounting for 11.5 per cent of Myanmar's total FDI. China emerged as Myanmar's second largest trading partner and largest investor. A hallmark of Chinese investment in Myanmar is a heavy concentration in strategic sectors. In his June 2010 visit, Premier Wen Jiabao signed 15 agreements on economic cooperation, including investments in a natural gas pipeline and hydropower station, and pledges for expanded development assistance (*The New Nation*, 22 September 2010).

In 1991, China pledged its first major aid grant of 50 million CNY (US$8.6 million) to Myanmar. In July 1993, China promised an interest-free loan of 50 million CNY (US$8.6 million), and the two sides celebrated the opening of the Yangon–Thanlyin Bridge, supported by loans of 169 million CNY (US$29.1 million) (Aung Myoe, 2007, p. 22). Between 1997 and 2006, China provided some 200 million CNY (US$24.2

Table 1. Chinese ODA to Myanmar (US$ million)

Year	Grant	Loan	Debt Relief
1997	–	12	0.6
1998	–	–	–
1999	–	6	–
2000	–	9.6	–
2001	–	13	–
2002	–	12	–
2003	6	200	0.6
2004	9.7	30	–
2005	–	–	–
2006	8.8	200	–

Source: Aung Myoe, 2007, p. 20.

million) in grants, 685 million CNY (US$82.7 million) and US$400 million in subsidised loans, and 10 million CNY (US$1.2 million) in debt relief (see Table 1). Recent projects include the rail, road and bridge project linking Yangon and Thanlyin, communication technology, agricultural machinery, hydropower projects, and construction of stadiums and government buildings. China also supported a number of build, operate and transfer agreements, including the Hmawbi Rubber Ball Factory, Tyre Factory, Belin Sugar Mill, Shwedaung Textile Mill and Meikhtila Textile Mill (Aung Myoe, 2007, p. 31).

In January 2003, Beijing pledged a grant of 50 million CNY (US$6 million) to support agricultural projects and three small-scale hydroelectric plants, and a loan of US $200 million to support construction of the Yeywa hydropower plant. In February 2006, China provided another US$200 million loan, this time for the procurement of drilling materials for oil drilling rigs and the construction of a urea fertiliser plant. China's support for state-owned factories in Myanmar is often followed by substantial orders for equipment supplied by Chinese firms. For instance, the 2004 loan package of 80 million CNY (US$9.7 million) included 15 million CNY (US$1.8 million) for Myanmar's Ministry of Industry to build a new generator plant at Sinde. The Ministry then purchased generator parts for this plant from China National Machinery Import and Export Corporation. Soon after this grant, the China-owned XJ Group Corporation and Henan Diesel Engine Group became a key player in building the Ministry's multipurpose diesel engine plant in Thagara, a project worth a potential US$112 million (Aung Myoe, 2007, p. 38). A 2007 study by the Burma Project identified 26 Chinese multinational corporations involved in 62 hydropower, oil, gas and mining projects in Burma (Zhao, 2008, p. 212).

China has expanded its support for infrastructure projects on boat, rail and road links, enabling goods to be shipped in by sea to container ports in Myanmar, offloaded, and then taken by rail and river barge over to the Chinese border. For instance, local county governments in Yunnan Province built the 95-kilometre-long Tengchong–Myitkyina road at a cost of 192 million CNY (US$23.2 million) and upgraded the Zhangfeng–Bhamo road at a cost of 28 million CNY (US$3.38 million), also agreeing to cover all construction and maintenance costs for a decade before turning the road maintenance over to Myanmar (Aung Myoe, 2007, p. 40). These highways will be valuable additions to Myanmar's infrastructure, aiding its ability to develop its industrial sector and market its agricultural products. The links also offer China the potential to circumvent the Malacca Strait by shipping oil and other resources directly across Myanmar. With over 80 per cent of China's oil imports passing through the Malacca Strait and with its growing dependence on imported oil, China seeks to reduce its vulnerability to US control of the Malacca Strait. In November 2009, China began to build a crude oil port on Maday Island on Myanmar's western coast as part of a crude oil pipeline scheme connecting the port with the Chinese border town of Ruili. Supported by some US$2.5 billion of Chinese funding, the scheme is designed to carry 12 million tonnes of crude oil a year (Pipelines, 2009, p. 14).

China's support for hydropower also continued to develop – three controversial dams on the Salween River in Myanmar apparently all received Chinese financial support – the electricity from which is likely to find its way back to China's Yunnan province (Middleton, 2007). Such dams play a critical role in achieving China's goal of generating 15 per cent of its power from non-fossil-fuel sources by 2020, up from about 8 per cent now. Building dams "is an important step taken by the Chinese Government to vigorously develop renewable and clean energy and contribute to the global endeavour to counter climate change," as Song Tao, China's Vice-Minister of Foreign Affairs told the Mekong River Commission in April 2010 (Lee, 2010).

China's assistance program to Myanmar has grown in concert with its strategic investments. In response to Cyclone Nargis in May 2008, China provided US$5 million in humanitarian aid, delivered through bilateral channels rather than via the United Nations' combined relief appeal (Bezlova, 2008). In April 2009, China announced a massive aid package to help Southeast Asian countries cope with the global financial crisis. The offer included a US$10 billion contribution to initiate a China–ASEAN Investment Cooperation Fund, geared for cooperation in infrastructure construction, energy and natural resources development, and improvements in information and communications; a US$15 billion line of commercial credit targeted at poorer ASEAN states (including US$1.7 billion for cooperative projects); and an additional US$39.7 million for "special aid" to Cambodia, Laos and Myanmar, US$5 million to the China–ASEAN Cooperation Fund, and US$900,000 to the ASEAN plus China, Japan and South Korea Cooperation Fund (McCartan, 2009).

China's aid and investment package builds upon its support for the Asian Development Bank (ADB)'s Greater Mekong Sub-region (GMS) program. Since 1992, the GMS program has promoted economic and social development via economic linkages and regional cooperation among the six Mekong countries. China has been particularly enthusiastic about the GMS Economic Corridors project, which includes a network of roads connecting all countries in the GMS including China's Yunnan Province. Further connectivity is extended through an expanding regional power grid and planned railway links from Kunming to Singapore, funded in part through development assistance and investment from China (ADB, 2010, p. 60). China's support appears linked to Beijing's high-speed railway diplomacy – a grand vision for linking China's high-speed rail network to its Asian neighbours (Holslag, 2010, p. 653). China has also actively supported the Intergovernmental Agreement on Regional Power Trade (IGA), an agreement reached among six Mekong River countries in 2002 supported by the ADB and the World Bank in order to facilitate joint electricity generation and distribution throughout the Mekong region. Under the IGA framework, electricity generated in Myanmar and Laos first began to supply China in 2006 (Zha, 2011, p. 2).

Yet even as Beijing gained greater access to Myanmar's natural resources and strategic location, Chinese leaders began to face greater diplomatic frustration. After casting its first non-Taiwan-related veto in the Security Council since 1973 in order to defeat a US–UK sponsored Security Council draft resolution on the situation in Myanmar in 2007, Beijing began to proactively promote international engagement, urge economic reform, and broker negotiations with rebel groups (International Crisis Group, 2009, p. 4). As a *Foreign Affairs* article notes:

> China's patience with the Burmese junta has been wearing thin recently. For several years, Beijing encouraged it to undertake Chinese-style economic and political reforms in order to help the regime consolidate its rule, ensure stability, and regain international acceptability. It supported former Prime Minister Khin Nyunt, whom it considered a Deng-style reformist – only to see him ousted in 2004. As the Burmese regime hardened further, China's confidence in the junta's capacity or willingness to reform faded (Kleine-Albrandt and Small, 2008, pp. 48–49).

Tensions broke into the open in 2009, after the Burmese armed forces launched an offensive against a minority Kokang armed group, the Myanmar National Democratic Alliance, along the Chinese border, causing more than 37,000 refugees to flee to China. In response, Chinese Vice-President Xi Jinping travelled to Myanmar, informing Senior General Than Shwe that China expected to see "political stability, economic development and national reconciliation … so as to guarantee the stability in its border area with China" (Moe, 2009). A year later, the two sides appeared to have reached an implicit understanding: Myanmar would refrain from actions exacerbating instability along the border while China pledged not to accept and support any groups carrying out anti-Myanmar Government activities in its border areas (Roughneen, 2010, p. 2).

Relations were further shaken by President Thein Sein's September 2011 announcement that he was suspending construction of the Myitsone dam for the duration of his term in office – until 2015. Of a series of seven Chinese-built dams planned on the Irrawaddy, the Myitsone was to be the largest, and at about 150 metres, one of the highest in the world. Reportedly, 90 per cent of the electricity generated would be exported back to China, while the dam's reservoir would flood an area the size of Singapore, driving over 10,000 people, mainly from the Kachin ethnic group, from their ancestral lands. This area straddles territory controlled by the Kachin Independence Organisation (KIO). After the KIO's May 2011 warning that building the dam would lead to "civil war", fighting between government forces and the KIO's armed wing increased sharply, forcing thousands of villagers to flee the area (Ruo, 2011). Indeed, President Thein Sein's justification for suspending the project was that it was "contrary to the will of the people". As Jacqueline Menager (2011, p. 2) notes, "The suspension represents two important developments: the government responding to public opinion, something previously unheard of; and a challenge to Chinese dominance in Burma".

In response, the president of China Power Investment Corporation (CPI), the dam's main investor, pronounced himself "totally astonished" by the decision and hinted at "a series of legal issues" (Kaung, 2011). China's Foreign Ministry spokesperson called for

Myanmar "to guarantee the lawful and legitimate rights and interests of Chinese companies", urging both sides to "properly handle" the matter through "friendly consultations" (Jackson, 2011). Chinese media detected Western influence behind the decision, arguing that Thein Sein used the stoppage to send a credible sign to the West about the seriousness of his domestic political reform agenda (Gui, 2011). The nationalistic newspaper *Global Times* sounded a more ominous tone, warning on 24 October 2011 that "the whimsical change will inevitably send troubling signals to potential investors". Yet as China's diplomatic frustration mounted, Japan seized upon domestic shifts within Myanmar to restore its economic engagement with the country – with foreign aid once again playing a central role.

Japan's Return to Myanmar

Seeking greater engagement but limited by diplomatic and legal constraints, Japan in recent years has sought to embed cooperation with Myanmar within a regional framework oriented around the Mekong River countries. In December 2006, Tokyo established the Mekong–Japan Partnership Program, including Myanmar along with Cambodia, Lao PDR, Thailand and Vietnam. In February 2009, Japanese Prime Minister Hatoyama Yukio travelled to Yangon, Myanmar's former capital, to launch the "Japan–Mekong exchange year", pledging US$5.6 billion in aid over the coming three years for the five Mekong River nations. Eighty per cent of the funding was dedicated to low-interest yen loans for projects ranging from regional highway links to water projects and technological training. The aid came on top of Japan's previous pledges of US$4.5 billion in aid for the region since 2007 (Makino, 2009).

In 2010, Myanmar embarked upon a series of dramatic political reforms that Emma Larkin (2012) dubs the "Burmese Spring", including the loosening of political and economic controls and releasing Aung San Suu Kyi from house arrest. In response, Tokyo began to ramp up its economic assistance. In September 2011, Japan pledged some 800 million yen (US$10 million) in grant aid. On the eve of historic Parliamentary by-elections, in March 2012 Tokyo promised another 1.6 billion yen (US$2 million). Following Suu Kyi's election to Parliament, Gemba Koichiro, Japan's Foreign Minister, praised the elections as "a significant step in the democratisation of Myanmar" (Statement, 2012). A few weeks later, Prime Minister Noda Yoshihiko received Myanmar President Thein Sein in Tokyo, where he pledged an additional 1.65 billion yen (US $2.1 million) in grant aid. Noda also announced that Japan would forgive some 300 billion yen (US$3.7 billion) of Myanmar's debt, clearing the way for the resumption of ODA loans to Myanmar. The two sides agreed that Japanese funding would support infrastructure construction in Myanmar's planned Thilawa Special Economic Zone, particularly its port facilities. The announcement came on the heels of the fourth Japan–Mekong Summit, in which Japan pledged to provide 600 billion yen (US$7.4 billion) in development aid and expertise to Myanmar, along with Thailand, Vietnam, Cambodia and Laos (see http://www.mofa.go.jp/region/asia-paci/myanmar/thein_sein_1204/index.html). Once again, Tokyo used its aid program to signal its enthusiasm for broader economic and diplomatic engagement with Myanmar.

China and Japan in Myanmar – Competition or Cooperation?

Both China and Japan have repeatedly used their foreign aid programs to advance broader strategic and economic objectives in Asia, including securing access to strategic natural resources. For much of the post-WWII period, Japan's aid program was utilised to bolster market access for Japanese firms, expand exports, and gain access to key natural resources in developing countries. China, one of the largest recipients of Japanese aid, followed a similar approach in its revamped aid program from the mid-1990s. These strategies have been evident in their respective aid programs in Myanmar. Shifts in Myanmar's domestic political environment have provided critical openings for both countries to expand their aid programs and promote their economic interests in Myanmar – China after 1988, and Japan since 2010. While the recent political opening in Myanmar has enabled Japan to resume its economic engagement, with aid once again leading the way, China's relations with Myanmar are facing greater difficulties as the domestic and diplomatic environment becomes more crowded and complex.

Given the strategic significance of Myanmar, there remains a substantial risk that China and Japan will end up competing over influence and access to national resources. Institutionalised cooperation is essential for minimising bilateral tensions between the two Northeast Asian powers. The annual Japan–China Mekong Policy Dialogue Forum, established in 2008, provides a venue for information sharing and discussion on joint projects in the region. Yet cooperation to date has remained limited to formal meetings between aid officials. More could be done on both sides to enhance information sharing, facilitate visits to each other's aid projects, incorporate local partners into three-way projects, and establish collaborative training programs. Given the parallels in their development trajectories and the similar histories of their aid programs, such collaboration would foster trust between China and Japan while enhancing the effectiveness of their development efforts in Myanmar. Ultimately, pragmatism offers the best hope. It is in both Tokyo's and Beijing's interests to find ways to ensure that they avoid ending up in a costly bidding war for influence and resources within Myanmar, and throughout the Mekong region.

Acknowledgments

An earlier version of this article was first presented at the 'International Politics of Resources' Research Workshop, University of Technology Sydney on 28–29 July 2011. The author expresses appreciation to the workshop organisers and participants for their helpful comments and suggestions.

References

Aquino, Carlos H., Jr. and Josephine Jensen-Joson (2009) *China's official development assistance to the Philippines* (Manila: The Official Development Assistance Watch – Philippines).

Arase, David (1995) *Buying power: The political economy of Japan's foreign aid* (Boulder: Lynne Reinner Publishers).

Asian Development Bank (2010) Toward sustainable and balanced development: Strategy and action plan for the greater Mekong sub-region north–south economic corridor (Manila: ADB). Available at http://www.adb.org/sites/default/files/pub/2010/gms-north-south-action-plan.pdf, accessed 9 December 2012.

Aung Myoe (2007) *Sino-Myanmar economic relations since 1988*. Asia Research Institute Working Paper No. 86 (Singapore: National University of Singapore).

Bezlova, Antoaneta (2008) Earthquake lets China off the hook. *Asia Times* 28, May. Available at http://www.atimes.com/atimes/China/JE28Ad01.html, accessed 10 October 2010.

Bin Ke (2008) Yi li xuanzhe yu zhongguo duiwai yuanzu de bianhua [The choice between justice and virtue, and changes in China's foreign aid]. *Xiangchao* [Hunan Tide], August, p. 36.

Bräutigam, Deborah (2009) *The dragon's gift: The real story of China in Africa* (Oxford: Oxford University Press).

Bräutigam, Deborah (2010) *China, Africa and the international aid architecture*. Working Paper for the African Development Bank 107, April.

Brooks, William L. and Robert M. Orr, Jr. (1985) Japan's foreign economic assistance. *Asian Survey* 25(3), pp. 322–40.

Edström, Bert (2009) *Japan and the Myanmar conundrum*. Institute for Security and Development Policy. Available at http://pwvb.rit-alumni.info/japan_and_the_myanmar_conundrum_-_final.pdf, accessed 10 September 2011.

Ensign, Margee M. (1992) *Doing good or doing well: Japan's foreign aid program* (New York: Columbia University Press).

Export–Import Bank of China (2006) *Annual Report 2006*. Available at http://english.eximbank.gov.cn/annual/2006fm.shtml, accessed 5 September 2011.

Foster, Malcolm (2012) Japan to forgive $3.7 billion of Myanmar's debt. *Seattle Times*, 22 April. Available at http://seattletimes.nwsource.com/avantgo/2018035400.html, accessed 11 June 2012.

Government of Japan, Ministry of Foreign Affairs (1992) *Japan's official development assistance, 1992* (Tokyo: Association for Promotion of International Cooperation).

Gui Qiangli (2011) Miandian jiaoting shuiba gongcheng de Beijing: xifang tiaofa zhongmian guanxi [Behind the stoppage of the Myitsone dam: The West poking at Sino-Myanmar relations]. *Caixun*, 4 October. Available at http://international.caixun.com/content/20111004/CX01ttu0.html, accessed 10 November 2011.

Holslag, Jonathan (2010) China's roads to influence. *Asian Survey* 50(4), pp. 641–62.

Hook, Steven W. and Guang Zhang (1998) Japan's aid policy since the Cold War: Rhetoric and reality. *Asian Survey* 38(11), pp. 1051–66.

International Crisis Group (2009) *China's Myanmar dilemma*. Asia Report No. 177.

Jackson, Allison (2011) China raps Myanmar over dam project. AFP, 1 October. Available at http://ph.news.yahoo.com/china-raps-myanmar-over-dam-project-085151191.html, accessed 23 November 2011.

Katada, Saori (2010) Old visions and new actors in foreign aid politics: Explaining changes in Japanese ODA policy to China, in Warren Leheny (ed.), *Japanese aid and the construction of global development: Inescapable solutions*, pp. 54–75 (London: Routledge).

Kaung, Ba (2011) Burma dam decision "bewildering": CPI president. *The Irrawaddy*, 4 October. Available at http://www.irrawaddy.org/article.php?art_id=22191, accessed 22 November 2011.

Kleine-Albrandt, Stephanie and Andrew Small (2008) China's new dictatorship diplomacy: Is Beijing parting with pariahs? *Foreign Affairs,* Jan/Feb, pp. 38–56.

Kobayashi, Takaaki (2008) *Evolution of China's aid policy*. Japan Bank for International Cooperation Working Paper No. 27. Available at http://www.jbic.go.jp/en/research/report/working-paper/pdf/wp27_e.pdf, accessed 10 September 2011.

Koppel, Bruce M. and Robert M. Orr (1993) A donor of consequence: Japan as a foreign aid power, in Bruce M. Koppel and Robert M. Orr (eds), *Japan's foreign aid: Power and policy in a new era*, pp. 3–26 (Boulder: Westview).

Krasner, Stephen D. (1978) *Defending the national interest: Raw materials investments and U.S. foreign policy* (Princeton: Princeton University Press).

Kreft, Heinrich (2006) China's quest for energy. *Hoover Policy Review* 139, Oct–Nov. Available at http://www.hoover.org/publications/policy-review/article/7941, accessed 20 November 2011.

Lancaster, Carol (2006) *Foreign aid: Diplomacy, development, domestic politics* (Chicago: University of Chicago Press).

Larkin, Emma (2012) The awakening: Inside the Burmese spring. *The New Republic*, 2 February. Available at http://www.tnr.com/article/world/magazine/99537/burma-spring-aung-san-suu-kyi, accessed 10 December 2012.

Lee, Yoolim (2010) China hydropower dams in Mekong River give shocks to 60 million. *Bloomberg Markets Magazine*, 26 October. Available at http://www.bloomberg.com/news/2010-10-26/china-hydropower-dams-in-mekong-river-give-shocks-to-60-million.html, accessed 23 November 2011.

Makino, Catherine (2009) Japan: Rivalry with China spurs hefty aid package to Mekong. *Inter Press Service*, 11 November.

McCartan, Brian (2010) A helping Chinese hand. *Asia Times*, 30 April. Available at http://www.atimes.com/atimes/Southeast_Asia/KD30Ae01.html, accessed 10 June 2010.

Menager, Jacqueline (2011) Aung San Suu Kyi and Burma's change of heart. *East Asia Forum*, 27 October. Available at http://www.eastasiaforum.org/2011/10/27/aung-san-suu-kyi-and-burma-s-change-of-heart/, accessed 10 September 2011.

Middleton, Carl (2007) Building friendships, building dams. *World Rivers Review*, 17 September. Available at http://www.internationalrivers.org/southeast-asia/building-friendships-building-dams, accessed 20 November 2011.

Ministry of Commerce, China (2007) *Basic state of China's external assistance 2006*. Available at http://jjhzj.mofcom.gov.cn/aaarticle/g/200701043l6858.html, accessed 10 September 2011.

Moe, Wai (2009) China tells Than Shwe it wants stability, national reconciliation. *The Irrawaddy*, 21 December. Available at http://www.irrawaddy.org/article.php?art_id=17439, accessed 22 November 2011.

Oishi, Mikio and Fumitaka Furuoka (2003) Can Japanese aid be an effective tool of influence? Case studies of Cambodia and Burma. *Asian Survey* 14(6), pp. 890–907.

Okita, Saburo (1974) Natural resource dependency and Japanese foreign policy. *Foreign Affairs* 52(4), July, pp. 714–24.

Ott, Marvin (1985) Japan's growing involvement in the Middle East. *SAIS Review*, Summer-Fall, pp. 221–32.

Pipelines in the Pipeline (2010) PetroMin Pipeliner, October–December, pp. 14–20. Available at http://www.safan.com/safannews/pdf/pin301210.pdf, accessed 9 December 2012.

Potter, David Matthew (1992) Japan's foreign aid to Thailand and the Philippines. Unpublished PhD dissertation, University of California, Santa Barbara.

Rix, Alan (1980) *Japan's economic aid: Policy-making and politics* (New York: St Martin's Press).

Roughneen, Simon (2010) Burma finds protective brother in China. *ISN Security Watch*, 14 September.

Ruo, Liao (2011) Lessons from the Irrawaddy. *China Dialogue*, 10 October. Available at http://www.chinadialogue.net/article/show/single/en/4574-Lessons-from-the-Irrawaddy, accessed 21 August 2011.

Shang Changfeng (2010) Wenge shiqi zhongguode duiwai yuanzhu [Chinese foreign aid in the era of the cultural revolution]. *Dangshi wenhui* [Party History Digest] 2, pp. 56–57.

Söderberg, Marie (2002) *Changes in Japanese foreign aid policy*. Stockholm School of Economics Working Paper 157.

Statement by the Minister for Foreign Affairs of Japan on the Legislature By-Election in Myanmar (4 April 2012). Available at http://www.mofa.go.jp/announce/announce/2012/4/0404_01.html, accessed 11 June 2012.

Streeford, Patrick (2007) Japanese ODA diplomacy towards Myanmar: A test for the ODA charter. *Ritsumeikan Annual Review of International Studies* 6, pp. 65–77.

Sun Luxi (2007) Cong guojia liyi shiyexia kan zhongguo jianguo yilai de duiwai yuanzhu zhengce [Viewing China's foreign aid policy since China's establishment from the perspective of national interests]. *Shidai jinrong* [Current Finance] 11(356), pp. 9–10.

Takamine, Tsukasa (2006) *Japan's development aid to China: The long-running foreign policy of engagement* (New York: Routledge).

Wan, Ming (2006) *Sino-Japanese relations: Interaction, logic, and transformation* (Stanford: Stanford University Press).

Yang Hongxi (2009) Zhongguo duiwai yuanzhu de huigu yu fazhan [Reflections and development of China's foreign aid]. *Xuexi yukan* [Study Times] 439, November, pp. 40–42.

Yang Fangyuan (2009) Lijie zhongguo ruanshili de sange weidu: wenhua waijiao, duobian waijioa, duiwai yuanzhuzhangce [Understanding the three dimensions of Chinese soft power: Cultural diplomacy, multilateral diplomacy, and foreign aid policy]. Consumer Guide [*Xiaofei daokan*], August, p. 231.

Zha, Daojiong (2011) China and its southern neighbours: Issues in power connectivity. *RSIS Commentary* 147, 14 October.

Zhai Dongsheng (2010) Duiwai yuanzhu de shenzhen daode yuanze [The cautious morality principles of foreign aid]. *CEOCIO*, 5 April, p. 92.

Zhang Guangrong (2010) Zhongguo de ziyuan nengyuanlei jingwai touzi jiben wenti yanjiu [Research on several basic issues of China's resource and energy overseas investment]. (Beijing: China Economic Publishing House).

Zhao, Suisheng (2008) China's global search for energy security: Cooperation and competition in the Asia-Pacific. *Journal of Contemporary China* 17(55), pp. 207–27.

Zweig, David and Jianbai Bi (2005) China's global hunt for energy. *Foreign Affairs* 84(5), pp. 25–38.

Chinese Aid in the South Pacific: Linked to Resources?

PHILIPPA BRANT

Lowy Institute for International Policy

Abstract: *China's emergence as a global development actor has implications for developing countries and "traditional" donor agencies. Its current provision of foreign aid and other forms of development assistance to developing countries throughout the world presents both opportunities and challenges for all actors. At the same time, China's growing need for natural resources and its policy of securing access through state-led "resource diplomacy" are causing concern. While most scholars and commentators are focused on the "China in Africa" dimension, China's engagement in the South Pacific region has also been growing rapidly over the past decade and offers some interesting and unique insights. This article examines the dynamics of China's provision of foreign aid and its quest for natural resources in the South Pacific region, with comparative references to other regions. Drawing particularly upon interviews and site visits in Fiji and Papua New Guinea, it argues that although major commercial resource contracts do appear to be supported by Chinese Government assistance, resources deals are not explicitly part of Chinese foreign aid in the region.*

Introduction

The emergence of China as a significant provider of development assistance operating outside the dominant aid system has prompted heightened interest within academic, public and policymaking circles. This increased presence in many developing countries is changing the dynamics of "development" and foreign aid provision in ways that are really only beginning to be seen and understood. At the same time, there has been increased concern about the implications of China's growing desire for natural resources and its efforts to secure access to these through state-led "resource diplomacy". China's provision of development assistance is often seen to be an important part of this strategy.

China's quest for energy resources to fuel its ever-growing economic development has prompted it to expand and deepen its relationships with oil and energy producing nations around the globe. Beijing's "go out" strategy[1] that began in the mid-1990s is predicated on a strong collaboration between the state and Chinese companies, which have been encouraged through incentives and other forms of state support to invest in overseas markets. Although this strategy involves numerous sectors, international attention has focused primarily on the energy sector and the role of China's national oil companies (NOCs), which are often perceived as presenting a threat to Western interests. As Mikkal Herberg (2011) argues, "although closer analysis suggests the Beijing-NOC collaboration is far less strategic than it often appears and many investments are much more normally market-driven than state-driven, the process appears from the outside as opaque and strategic in intent". Since examples of the involvement of the Chinese state apparatus in securing energy deals are plentiful,[2] it is often assumed that Chinese Government foreign aid is also primarily directed towards accessing natural resources.

Chinese aid expert Deborah Brautigam in her discussion of the "myths" surrounding Chinese aid explores this misconception, which seems to be perpetuated not only by journalists but also by policymakers and development organisations. A World Bank study, for example, states that "most Chinese government funded projects in Sub-Saharan Africa are ultimately aimed at securing a flow of Sub-Saharan Africa's natural resources for export to China" (Foster et al., 2008, p. 64). Brautigam (2009, p. 278) argues that although it is easy to see how this belief has arisen, given China's very active involvement in resource-rich areas, "the notion that aid is offered mainly as a quid pro quo exchange for resources ignores several facts", including the fairly even distribution of grants and zero-interest loans across resource and non-resource rich countries.

Discussion of the nexus between foreign aid and natural resource acquisition in China's engagement with other developing countries has in general drawn upon examples of oil and mining in Africa, mining and agribusiness in Latin America, and hydropower in the Mekong region. China's involvement in the South Pacific region is logically much smaller than in other parts of the world, although its impact on Pacific Island nations is significant. Since the Pacific Island nations are traditionally regarded as being under the Western "sphere of influence", China's political and economic engagement over the last decade has caused concern amongst some policymakers and commentators, especially in Australia and New Zealand (see, for example, Commonwealth of Australia, 2009, pp. 44–46; Hanson, 2009). Although many of the countries are small island states, some are well endowed with minerals, timber and exclusive economic zones for fishing, and in addition have the rights to as-yet unexploited seabed resources (Porter and Wesley-Smith, 2010).

This article examines the dynamics of China's provision of foreign aid in the South Pacific and considers the connections with its quest for natural resources, drawing particularly upon interviews and site visits in Fiji and Papua New Guinea (PNG), undertaken in September–October 2009. It concludes that unlike examples in other regions, resource deals in the South Pacific are not explicitly part of Chinese aid. Chinese resource companies operating in the region are nevertheless state-owned enterprises (SOEs), and so the line between "commercial" activity and "state" involvement is often blurred, which can be seen on occasions when the Chinese Government comes to the aid of Chinese companies or utilises its development assistance to support SOEs' activities. Examples of such occurrences can offer further insight into the operation and utilisation of Chinese aid towards broader strategic goals.

This article first provides an analysis of Chinese foreign aid in general, including its relationship to resource acquisition. It then examines the South Pacific context, and China's engagement in the region, before presenting a more detailed discussion of the relationship between Chinese aid and investment and state and commercial endeavours. It concludes by arguing that Chinese aid in the South Pacific should not be viewed (and dismissed) as simply an element of Beijing's strategy to secure access to natural resources.

Chinese Foreign Aid

The impact of Chinese foreign aid is undoubtedly being felt by both developing countries and OECD–DAC donors alike. Yet, until recently at least, knowledge of the mechanisms, practices, and even the principles and objectives remained sketchy. Misconceptions have been fed in part by China's own lack of transparency and sensitivity in releasing details of its aid program, and in part by the desire of some Western commentators to use Chinese aid in support of their "China Threat" theses (for example, Windybank, 2005; Shie, 2007). This section provides a general overview of the core elements of Chinese foreign aid and the mechanisms that relate directly to resource acquisition.

Contrary to current rhetoric, China is not a "new" or "emerging" donor, since its history of aid giving goes back 60 years (Varrall, 2012, pp. 142–43). The Chinese Government refers to three stages of Chinese aid provision: 1950–78, 1978–mid 1990s, and 1990s onwards, with the changes reflecting shifts in China's own development situation and strategies (Ministry of Commerce, 2007). China started to emerge as a more significant donor in the late 1990s, a development that was related to its own domestic development policy of "going out", and to its greater involvement in multilateral and international organisations. 2004 saw the beginning of yet another "new stage" in China's foreign aid program, marked by a massive and sustained increase in levels of aid: between 2004 and 2009 China's foreign aid budget increased by an average of 29.4 per cent per year (People's Republic of China, 2011) and in 2011 totalled approximately US$4.5 billion (Interview CN064, 6 May 2011).

The Chinese Government, in its first ever White Paper on Foreign Aid (referred to hereafter as the White Paper), released in April 2011, frames China's provision of aid as operating within the context of China's position as a developing country but also as part of the fulfilment of its international responsibilities. In what can perhaps be regarded as the stated overall objective, China says it is providing foreign aid to "help recipient countries to strengthen their self-development capacity, enrich and improve their peoples' livelihood, and promote their economic growth and social progress" (People's Republic of China, 2011). There is a clear declaration that Chinese aid is a "model with its own characteristics", though no clear declaration of what this actually means (see Brant, 2011b for an overview of the White Paper).

Table 1. Comparisons of Aid Calculations

	China	DAC
Military aid	√	×
Scholarships	×	√
Debt relief	×	√
Loans for joint-venture projects	√	×
Concessional loans	% under market rate	Entire face value
Newly settled refugees	×	√ Some costs

One of the challenges in understanding Chinese aid is that there is no statement defining what China considers and calculates as "aid". In addition, although the terminology and details in the White Paper seem to be comparable to the OECD–DAC definitions, there are some significant differences that make Chinese aid directly incomparable with DAC aid data (see Table 1).

Examining official Chinese statements and policy documents tells us about the way China *wants* its aid to be perceived. Interestingly, at an official exhibition held in Beijing in 2010 to commemorate the 60th anniversary of Chinese foreign aid, China used the English term "aid". Until this point, China had preferred to use broader terms such as "economic cooperation" and "development assistance" when referring to its engagement with other developing countries. The common narrative of Chinese aid is to take it back to the "Eight Principles for China's Aid to Foreign Countries" announced by Premier Zhou Enlai in 1964, which have been rearticulated by successive Chinese leaders and officials (People's Republic of China, 2003). China's White Paper builds upon this and articulates five basic features underlying Chinese aid policy:

- Unremittingly helping recipient countries build up their self-development capacity
- Imposing no political conditions … respect recipient countries' right to select their own path and model of development
- Adhering to equality, mutual benefit and common development
- Remaining realistic while striving for the best
- Keeping pace with the times and paying attention to reform and innovation.

China frames the provision of aid as operating within the context of its position as a developing country and as "fall[ing] into the category of South–South Cooperation" (People's Republic of China, 2011). Chinese officials frequently embrace the idea that China is also a developing country with a shared history of (negative) experience with imperial powers. It can be a politically powerful message vis-à-vis the history of traditional donors' programs in many places. Carol Lancaster (2007, p. 5) argued in 2007 that

> for political reasons they want to project their own distinctive image in Asia, Africa, and Latin America – one of South–South cooperation … one of having emerged rapidly (but not yet completely) from those problems, and one that will provide them with a separate and privileged relationship with the governments they are helping and cultivating.

Yet the reality is more complex. Strauss (2009, p. 792) writes that even though

> official and semi-official speeches continue to be topped and tailed with the invocation of analogous pasts of underdevelopment, suffering at the hands of imperialism, and China's presumptive superiority in being an "all-weather" friend … [this] is now looking more than a little threadbare in the light of China's actual involvement … and its ever-increasing global presence.

In this sense, China is arguably no different from other donors in its disjuncture between its policy rhetoric and reality.

For both domestic and international reasons, and unlike "traditional donors", Chinese foreign aid does not include a focus on democracy, good governance or human rights, which has led to much criticism and concern in the West (see, for example, Naím, 2007). It instead stresses the importance of stimulating economic growth and implementing a development model based on each country's specific requirements and circumstances (Interview SP011, 17 September 2009). In this way China seeks to differentiate itself from traditional donors (Li, 2008).

Chinese foreign aid is provided in three forms: grants and interest-free loans (through state finances) and concessional loans administered through China Eximbank. According to the White Paper (2011), to the end of 2009 China had provided a total of 256.29 billion RMB (US$39.3 billion) in aid, made up of approximately 41 per cent in the form of grants, 30 per cent as interest-free loans and 29 per cent as concessional loans. A key problem in the literature on Chinese aid is the tendency to lump all of the different types of Chinese economic engagement together under the label "aid". This ambiguity is partly the result of China's lack of transparency in the types and terms and conditions of its assistance, but it is partly the result of lazy or confused analysis. The influential Congressional Research Service report 'China's Foreign Aid Activities in Africa, Latin America and Southeast Asia' (Lum et al., 2009), for example, includes FDI in its calculations and relies on media reports for its data (Interview US057, 19 April 2010). Despite these flaws, the data continue to be used by high-profile publications and reports (such as *The Economist*, 13 August 2011).

In terms of the forms of aid, China classifies activities into eight categories: complete projects; goods and materials; technical cooperation; human resource development cooperation; medical teams sent abroad; emergency humanitarian aid; volunteer programs in foreign countries; and debt relief. At present, 40 per cent of China's foreign aid expenditure is in the form of "complete projects".[3]

China's approach to providing foreign aid is still very much focused at the government-to-government level, with the "state" still an integral component in its engagement. This political, state-led strategy is fraught with increasing challenges, including in the practical implementation of projects and (lack of) relationships with civil society and local communities. For example, officials of recipient governments find it frustrating that decisions (even seemingly insignificant ones) need to be referred to Beijing for approval, and that the Chinese insist on going through foreign affairs departments rather than working with the relevant executing division (Interview SP032, 8 October 2009). In addition, as one representative from an international civil society organisation in Suva stated: "Chinese assistance is government-to-government; there is no support for NGOs. [This] shows that their engagement is political" (Interview SP022, 29 September 2009). In addition, the plethora of Chinese actors – the Government, large state-owned enterprises, provincial-level companies, small-scale entrepreneurs, and migrants (legal and illegal) – makes controlling its state-driven agenda increasingly challenging for the Chinese Government. These issues have been discussed particularly in the African context (for example, Gill and Reilly, 2007; Dobler, 2009) and they generally apply to the South Pacific region as well. Alden and Hughes (2009, p. 564) are correct in arguing that "when analysing a relationship like that between China and Africa, the tendency to take the state as the main unit of analysis can rightly be criticised for neglecting that 'China' is anything but a unitary actor". This is a particularly pertinent point to remember when analysing China's (state and commercial) involvement in the resources sector.

Aid for Resources?

The Chinese Government (along with many developing country leaders) stresses the concepts of "win-win", "equal partnership" and "mutual benefit" in the objectives and operation of Chinese aid. In reality, like most countries, China's provision of foreign aid is utilised to further its foreign policy objectives. Chinese foreign policy expert Linda Jakobson (2011) explains that China's "core interests" boil down to concerns about sovereignty, security and development. She cites high-ranking Chinese foreign policy official, Dai Bingguo, elaborating upon these as: China's political stability (stability of the CCP leadership and socialist system); sovereign security; and China's sustainable economic and social development. Chinese scholars such as Ding (2008, p. 195) concur, explaining that the official basic goals of China's foreign policy are to "preserve China's independence, sovereignty and territorial integrity", and to "create a favourable international environment for China's reform and opening up and modernisation construction". In terms of more specific objectives of foreign aid, these can be broadly defined as serving economic growth and development (both in other developing countries *and* within China), geopolitical imperatives, and China's desire to be seen as a responsible actor.[4]

The program for each recipient country features different elements of these objectives according to variations in the local situation and what each country has to offer China in return. As Takaaki Kobayashi (2008, p. 36) explains in a report for JBIC (Japan Bank for International Cooperation), "Chinese aid follows the win-win principle and is given in 'exchange' for 'something' that contributes to its national interest. This 'something' may change in different times and with different countries".

One of the most cited benefits of China's engagement is that "aid" is just a small part of a broader economic relationship. Its aid program is primarily run through MOFCOM, the Ministry of Commerce, and the Government draws upon a range of financial mechanisms ("aid" and "non-aid") when formulating economic and development agreements with other developing countries. As Brautigam (2010, p. 19) explains,

> In 2006 the Chinese Eximbank announced that it had developed a "package financing mode" that can combine lines of export buyer's credit (given to a borrowing country), export seller's credit (short term credits given to a Chinese company) and concessional loans (foreign aid) which can be offered together, sometimes, but not always, for a specific project.

With this background in mind, it should not be surprising that foreign aid is sometimes used as leverage or an entry point for longer-term economic engagement – an avenue of enquiry that I explore later in this paper.

As highlighted in the introduction, there is widespread belief that much Chinese aid is linked to resource acquisition. In what seems to be an attempt to answer this accusation, the White Paper contains the following passage:

> Of China's concessional loans, 61% are used to help developing countries to construct transportation, communications and electricity infrastructure, and 8.9% are used to support the development of energy and resources such as oil and minerals (People's Republic of China, 2011).

That said, the so-called "natural resource-backed loans and lines of credit" are of particular interest to us here as this seems to be the form of development assistance utilised directly for access to resources. Brautigam explains:

> A country uses its natural resources to attract and guarantee an infrastructure loan from China on better commercial terms than it is likely to get from commercial banks. The loan is used to build infrastructure... In some cases ... existing natural resource exports are used as *security to guarantee repayment*. In other cases, the loan will be contingent on a Chinese company gaining preferential access to a block of natural resources that will be developed, and the proceeds used to repay the loan (Brautigam, 2010, p. 21, emphasis in original).

Known as the "Angola Mode" for its early use by China in Angola, this type of financing arrangement has caused much consternation amongst other donors and governments (see, for example, Jansson, 2011) despite having been used by donors giving aid *to* China in the 1980s. Although they cannot be considered "foreign aid" by DAC standards, these resource-secured loans can potentially be a useful instrument for development, but this appears to be contingent on the strength of local agency (Vines, 2011). Further discussion and assessment of this form of development assistance is beyond the scope of this paper, but awareness of this "mode" is important when examining Chinese engagement in the South Pacific.

Chinese Aid in the South Pacific

In discussions of Chinese foreign aid and engagement with developing countries around the world there is frequently very little (or no) mention of the South Pacific region or individual Pacific Island states. Yet China's role as a donor in the South Pacific region has been growing rapidly over the past decade and offers some interesting and unique insights into the links between aid, resources and investment. The role of Taiwan in the region also adds further complexities to China's involvement and the broader regional dynamics. After situating China's engagement in the broader South Pacific context, this section examines the debates about China's objectives, perceptions of its aid, and the main focus, types and modes of Chinese aid in the region.

The past decade has seen an increased Chinese presence in the South Pacific. Trade and investment have grown as China seeks new markets and access to resources such as timber, copper and palm oil. China's trade volume with Pacific countries reached a new high of US$3.66 billion in 2010, which is 50 per cent higher than the figure in the previous year (Han, 2011).

The China–Taiwan dynamic is significant in the South Pacific. Of the 14 countries in the region (excluding Australia and New Zealand), eight have diplomatic relations with China and six recognise Taiwan. The region has been the site of very active "chequebook diplomacy" in the past decade – including cases of both sides interfering in domestic politics (Crocombe, 2007, pp. 255–67). Pacific Island leaders, too, have been involved in playing China and Taiwan off against one another. There has recently been a tacit "agreement" between China and Taiwan not to actively compete for each other's allies[5] and it seems to be holding sway in the region, although according to those closely involved, the dynamics are still very interesting in regional forums (Interview SP004, 10 September 2009).

There is a lot of debate about why China is giving aid to the region. The less-than-transparent nature of China's foreign policy decisions and, in particular, its foreign aid programs, has resulted in the tendency to jump to negative conclusions about China's involvement and the "threat" it may pose to the region. Whilst concerns about Chinese practices are in many ways legitimate, this form of analysis has the tendency to ignore the role of Pacific Island countries themselves in determining their partners as well as the appeal of China vis-à-vis existing practices in the region.

Until recently, the dominant Western discourse has presented China's increased engagement in the South Pacific region as a "threat" to Western (in this case primarily Australia and New Zealand) strategic and development objectives, and has tended to characterise China as the "bully" or "dragon" of which South Pacific nations should be wary. Australia and New Zealand have traditionally seen the South Pacific as "their" backyard and over the past decade have focused particularly on promoting "good governance" and economic liberalisation through their aid programs. Australia has also played a significant role as a provider of security in response to political instability – the Regional Assistance Mission to the Solomon Islands is a prime example. The failed states discourse has been utilised to explain and in fact justify an increasingly interventionist approach to providing development assistance. In addition, Australia and New Zealand are currently trying to negotiate a regional free trade agreement known as PACER Plus, but many Pacific governments and NGOs are concerned about how the agreement will affect their economies and livelihoods (Maclellan, 2011; Morgan, 2012).

Following this strategic logic, Cold War-esque "China Threat" Western analyses, such as that offered by Susan Windybank (2005) or Tamara Shie (2007), cast suspicion on China's involvement, suggesting that China has devised a comprehensive strategy to take advantage of waning US interest in the region, and that it is attempting to penetrate the region through emigration, all of which has a negative impact on the "arc of instability" in the Pacific. Even though the academic debate has become more nuanced, this discourse remains influential in policymaking and think-tank circles (see, for example, Radio Australia, 2011b), and China's growing presence certainly appears to have sparked a renewed interest in the Pacific from traditional donors, including Japan, the EU and the US. For example, a representative from the Japanese Embassy in Fiji admitted that "Japan at the PALM 5 meeting increased its regional aid budget because of China. It saw what China last pledged and wanted to have a little more, so changed the budget" (Interview SP021, 28 September 2009).

The limited volume of analysis from Chinese scholars is generally at pains to point out that there is no evidence to suggest that China is trying to usurp Western predominance in the region. Jian Yang (2009), for example, concludes that the South Pacific is not essential to China's security strategy and that China has neither the hard power nor the soft power to become a hegemon in the region in the foreseeable future. He does note in his 2011 monograph, *The Pacific Islands in China's Grand Strategy*, that "China is playing an important role in the evolution of the regional order" (Yang, 2011, p. 145), but says its influence is not "deep rooted", and that it has relatively poor connections in non-economic areas. Yang and others such as Zhang (2007) acknowledge that China has a demand for the natural resources found in the region, but emphasise that China's engagement offers valuable development opportunities for the Pacific Island countries.

Chinese officials in the region employ familiar language when explaining China's engagement. Chinese Minister of Commerce Chen Deming stated at the China–PIC Forum in 2008:

> Politically, we believe every country, regardless of its size, strength and wealth is an equal member of the international community. We respect the individual choice of PICs for their way of development and their efforts in safeguarding sovereignty and regional stability (Chen, 2008).

Chinese Embassy officials also utilise these concepts when explaining China's approach to foreign aid. A Chinese Counsellor in Fiji stated: "China doesn't try to change recipient countries' systems. [It is] always based on equality and mutual respect" (Interview SP011, 17 September 2009).

There is no denying that Chinese aid in the region contains "national interest" objectives, as it does throughout the rest of the "developing world", but what the dominant analysis in reality objects to is not that China has a "strategic interest" in the region per se, but rather that it is potentially affecting *Western* "strategic interests". NGOs and aid practitioners from bilateral and multilateral agencies actually working in country are more willing to accept and articulate the hypocrisy of Western concerns about China's "strategic objectives" in the region, at least in private: "All donors have agendas – they differ in the extent they are hidden" (Interview SP013, 18 September 2009); "China is no different to other countries regarding wanting something from the region" (Interview SP017, 24 September 2009). As one donor representative admitted, "Western donors are just jealous and worried about increased competition" (Interview SP017, 24 September 2009). It seems that the more these concerns about China are articulated by Western officials and commentators the more they are dismissed by recipient communities as merely Western donors seeking to maintain "their" influence over their "special patch". Such hypocrisy is rarely lost on anyone, including the Chinese.

In terms of access to natural resources more specifically, the most comprehensive book to date on China's involvement in the region, *China in Oceania*, concludes that although China does have a significant interest in the resources in the region, "this is not sufficient to explain all aspects of its regional engagement" (Wesley-Smith and Porter, 2010, p. 22). As will be explored further below, Chinese aid is focused much more on other sectors. Despite this, the role of resources seems to be featuring more prominently in comments made by senior Western officials. New Zealand Foreign Minister McCully (2011), for example, recently stated:

> China is simply doing in our neighbourhood what it is doing in every neighbourhood around the globe: undertaking a level of engagement designed to secure access to resources on a scale that will meet its future needs, and establishing a presence through which it can make its other interests clear… Minerals, timber and fish are all commodities that the Pacific is able to trade, and that China wants to buy.

A more notable (and provocative) example is comments made by US Secretary of State Hillary Clinton to the Senate Foreign Relations Committee in March 2011:

> Let's just talk, you know, straight realpolitik. We are in a competition with China. Take Papua New Guinea: huge energy find... ExxonMobil is producing it. China is in there every day in every way, trying to figure out how it's going to come in behind us, come under us (Quinn, 2011).

Regardless of the reality, then, the perception remains of "China Energy Inc." at work and of China's use of the provision of development assistance in support of this.

Reliable figures on Chinese foreign aid to the region are difficult to confirm. Currently, the 'Economic and Technical Cooperation Framework'[6] outlined by Chinese Premier Wen Jiabao in 2006 still constitutes the guiding framework for the China–Pacific Islands relationship.[7] China's total aid (including concessional loans, grants and technical cooperation) approximates that of other important donors (Japan, NZ, European Commission), but is significantly less than Australia's. The most recent Lowy Institute report – 'China in the Pacific: The new banker in town' (Hanson and Fifita, 2011) – claims that Chinese aid to PNG accounted for 58 per cent of China's total aid to the region in 2009. This is unsurprising given PNG's growing economy, increased investment opportunities and natural resources vis-à-vis most of the other island states.

There is a general perception that Chinese foreign aid to the region has been increasing significantly during the past few years, with many donors and commentators particularly concerned about China's support for the Fijian Interim Government (Radio Australia, 2011b; Hanson and Hayward-Jones, 2009). The 2006 military coup in Fiji resulted in New Zealand, Australia and the EU (among others) suspending much of their development assistance to the Fijian Government. Although assistance to social sectors and poor communities continued – and is forecast to increase again, arguably in response to China's ongoing engagement – key donors such as Australia still do not provide any aid to or through Fijian Government authorities. At the same time, China has continued to strengthen its relationship with Fijian Government representatives, including Commodore Bainimarama.

China's foreign aid presence is certainly growing. However, despite reports suggesting more concessional loans, the April 2006 US$375 million Soft Loan Facility – a regional funding facility negotiated and disbursed bilaterally – currently constitutes the total pool of Chinese concessional loan funding to the region. Part of the confusion about annual Chinese aid figures, perhaps, stems from the fact that these funds have only recently begun to be accessed and disbursed. To compound this, political leaders and the media often tend to (re)announce the same funds numerous times – in part to trumpet China's support.

Some analysts and commentators present Chinese aid as appealing to recipients because of "negative" factors – lack of transparency, less-stringent environmental requirements, policy of "non-interference" and so forth (for the most prominent example, see Naím, 2007). Indeed, many aid practitioners and representatives from traditional donor agencies in Fiji and PNG presented Chinese aid in this way in interviews: "Many countries appreciate China because they give them what they want" (Interview SP010, 17 September 2009); "The total lack of conditionality of their aid is certainly appealing to any developing country" (Interview SP049, 20 October 2009). Whilst many of these concerns are undoubtedly legitimate, the attempt to paint China as "bad" vis-à-vis traditional donors and the suggestion that recipient governments only like China because of these characteristics entrenches the patronising approach to "developing" countries, which was a factor that led them to seek out China in the first place. It also ignores what may be real benefits of Chinese aid, and denies recipient governments and communities the ability not only to determine their own development partners, but to be able to assess for themselves the benefits and drawbacks of Chinese funding. As Wesley-Smith and Porter (2010, p. 2) note, "…while there is concern about such an enormous and relatively unfamiliar power acquiring a significant stake in Pacific futures, there is also clear appreciation for China's recent efforts in the region".

It is clear that China's aid programs are directed towards supporting China's national interests in some fashion, but in the Pacific at least, this does not take the form of being focused primarily on the natural resources sector. While the Chinese Government may not have clearly articulated "country strategies" for the region, Chinese officials in the Pacific have said that they want to diversify their development assistance and have identified four main areas of focus: infrastructure and public services; production and technical services; human resources and capacity enhancement; and regional funding (Interview SP011, 17 September 2009). As Ron Crocombe (2007, p. 218) correctly identified,

> China's aid moved to a new level in 2006 and China widened the range to a pattern more typical of the other major established donors. It provides for major increases in assistance to agriculture, education, forestry, fishing, tourism, manufacturing, telecommunications, aviation, shipping and health.

The following representative list of examples seeks to highlight the range of sectors towards which Chinese aid in the form of grants was directed over 2009–11.

- 50 training workshops in Fiji in 2010
- Provision of rice-growing experts, PNG
- US$800,000 for the Pacific Islands Forum regional development plan, 2009
- US$2.3 million to build the Navuso Bridge, Fiji
- On-going provision of medical supplies and doctors, Port Moresby General Hospital
- US$50,000 to Fiji Red Cross for Cyclone Tomas relief
- US$1 million for Kosrae coastal erosion, Federated States of Micronesia
- US$923,000 worth of military equipment to Tonga
- US$2.2 million worth of construction machines and equipment provided to Fiji

- A donation of US$50,000 to the PNG Red Cross to assist with the 2010 cholera outbreak.

In terms of concessional loans, a number of Chinese development assistance projects funded through the 2006 Soft Loan Facility have now (finally) been identified. The PNG Government identified two priority projects to be funded under the Chinese pledge: US$69.62 million for the Pacific Marine Industrial Zone Project (PMIZ) in Madang Province[8] and US$43.76 million for the University of Goroka Student Residence Building Project. In Fiji the Interim Government identified infrastructure, water supply and low-cost housing as projects to be funded under the scheme, including a request for the funds to be used for three roads to be built in 2011–12 (Tukuwasa, 2011). Tonga signed a US$45.28 million agreement with the Chinese, somewhat ironically, to fund the reconstruction after the (anti-Asian) riots that occurred in the capital in 2006, as well as US$42 million for roads. These examples suggest that the majority of Chinese concessional loan funding is still directed towards infrastructure-related projects. This is not surprising given it is an area that both China and Pacific Island Governments have identified as one where China can contribute, and the large size of the funding lends itself to larger infrastructural projects as well.

Chinese soft loans to the region are substantial vis-à-vis the relatively small national budgets and thus can have significant implications for Pacific Island countries and other donors. The benefit or success of Chinese loan assistance in many ways depends on the Pacific Island Governments' ability and willingness to select projects that will benefit their communities and to enforce their own requirements when negotiating terms and conditions.

Like other donors, China has issued extensive debt relief throughout the world and there is a pervasive perception amongst leaders in the region that Chinese loans will ultimately be forgiven. The loans provided by China to PNG in the 1980s and 1990s, for example, have yet to be repaid. A confidential Chinese Government document from 2008 that has been sighted by the author spells this out unambiguously:

> Despite the loan agreement stating that the Government of PNG should repay the loans, to date no principal payment has been made due to non-receipt of the billing advice, account details and other relevant information from Bank of China. We have made numerous requests to them to [provide] the payment advice albeit unsuccessfully.[9]

A PNG official confirmed this in an interview: "One of the difficulties is when we ask for information about when to pay, they don't provide it, so we say jokingly that it is actually a grant" (Interview SP027, 5 October 2009).

The issue of debt sustainability in the smaller island states has become a concern. In Tonga, for example, traditional lenders are concerned that the country has already reached a level of debt that is a problem, but it is negotiating further loans with China (with the expectation they will be forgiven down the track). Because of this concern, it is wary of providing any more loans due to repayment risks and is concerned that high levels of borrowing are undermining its own focus on improving financial management (Interview SP017, 24 September 2009). The Fijian Government recently asked China to write off some of the early loans, but the Chinese side refused for the time being (although it did not dismiss the option completely) (Interview SP017, 24 September 2009). One of the critiques of the concept of debt forgiveness is that recipient countries are often blamed for "irresponsible" borrowing, but the onus should also be on donors to ensure "responsible" lending. If China chooses to transform loans into grants, that is not a problem per se; however, what needs to be further examined is whether the perception that loans will ultimately be forgiven is prompting recipient governments to seek funding for projects that are not sustainable or are inappropriate. The US$13 million aquatic centre complex in Samoa is a pertinent example that has caused concern from traditional donors and the local community – about on-going maintenance costs and appropriateness of a "world class" facility in a city of fewer than 40,000 people (Lesa, 2009).

Significantly, the finance minister in the Cook Islands decided to put a Chinese loan "on hold" due to concerns about taking on more debt. The loan was for US$37 million (around 17 per cent of GDP!) for road and water projects. They also turned down a suggestion from CCECC (China Civil Engineering Construction Corporation) to ask the Chinese Embassy for funds to rebuild 200 homes destroyed by Cyclone Pat, in part because they want to engage as many locals as possible in the recovery work (Greig, 2010). This seems to indicate that the processes around Chinese aid are not static and in many respects depend upon the desires and strategies of recipient government leaders in negotiating good development outcomes for their communities.

Interviews did reveal that some recipient officials express frustration with Chinese aid, in part because "nothing is in writing". In discussions with the Chinese about terms and conditions of loans, including repayment periods and transformation into grants, "the Chinese always say, we can sort it out if and when it comes to that" (Interview SP026, 5 October 2009). Details of contracts and procurement are often scarce and recipient communities have legitimate issues about ensuring that the costs are accurate and they are getting best value for money. One provincial government official in PNG explained their concern:

> Say they have something worth 3 million Kina. With Australia you would know what it was all for. The Chinese always negotiate, try to bargain. And, even though it might be worth 3 million Kina, they have their own workers, materials, so what do we get out of it? (Interview SP044, 15 October 2009).

Another government official complained: "[We have] no way of verifying purchases and costs. They would say they had bought 'x' amount of concrete at 'x' cost, but [we] would never see any actual details" (Interview SP016, 24 September 2009).

This leads us to the issue of non-conditional aid. Alongside their insistence on respecting state sovereignty through non-interference, Chinese officials promote their norm of not attaching conditions to aid. Premier Wen at the inaugural China–PIC Forum in 2006 stated: "China stands ready to provide assistance without any *political* strings attached" (Wen, 2006, emphasis added). More recently, after the Pacific Islands Forum in 2009, Wang Yongqiu, head of Pacific relations in the Chinese Foreign Ministry said, "our aid comes with no political strings attached. It is not a means to exert political pressure or to seek political privileges..." (Callick, 2009). Significant is the emphasis on the lack of *political* conditions rather than conditionality per se. What is now well known is that Chinese foreign aid does have conditions attached, even if they are not political or policy-related. In the case of concessional loans in particular, the contractor must be a Chinese company and at least 50 per cent of the materials must be procured from China. Because of this feature, Chinese aid is heavily criticised by other donors (and recipient communities) for being "tied", against the OECD–DAC trend of "untying" aid. Interestingly, recipient government officials did not see this as a problem with Chinese aid per se. "In reality, most aid is still tied – to consultants and experts – so in this aspect China is no different" (Interview SP026, 5 October 2009). "Most Chinese aid is tied, like AusAID, to be frank" (Interview SP030, 7 October 2009).

The condition of using a Chinese contractor to implement an aid project has created numerous problems and highlights challenges within the Chinese aid mechanisms in the Pacific and elsewhere. For example, the Chinese Government, in outlining the process, explained that the recipient finds the contractor (Interview SP011, 17 September 2009). Technically, this may be true, but those involved in negotiating the housing loan deal in Fiji, for example, stressed that one of the biggest challenges was finding a Chinese company that was approved by the Chinese Government; a process fraught with issues and complicated by language barriers. The contract was advertised through the Fijian Embassy in Beijing and applicants applied directly to the relevant Fijian Government authorities. As someone closely involved in the negotiations explained: "[We] received a number of applications for the contract, would select one, only to find out they weren't approved... Even the Chinese Embassy here couldn't tell us if China Railway Group were approved" (Interview SP020, 28 September 2009).

A related issue throughout the region, which is especially explosive in PNG, is the influx of Chinese migrants who are brought in to work on Chinese aid (and investment) project sites. Interviews revealed that in the eyes of local communities there is no distinction between Chinese migrants who come "independently" and are involved in small-scale investments and those workers specifically brought in for Chinese projects. Whilst the issue is complex and problems are widespread there is a question increasingly being asked as to why the PNG Government, for example, "is bending the [immigration] rules for the Chinese" (Interview SP052, 21 October 2009). PNG has quite strong legal requirements regarding language skills and work permits, which are often being disregarded on both sides in order to facilitate Chinese funding (Smith, 2012).

This exploration of Chinese aid in the South Pacific has sought to highlight the dynamics of China's engagement, particularly the complexities relating to its objectives, priorities and financing modes. The following section will examine more specifically the nexus between Chinese aid and investment in natural resources in the region, offering some examples for comparison with other parts of the "developing world".

Chinese Aid: State and Commercial Interests

Western donors have been working for years to ensure a distinction between their commercial and aid endeavours in a country. "In the early years of official development assistance, donor countries commonly competed with each other in part by drawing on their ODA to subsidise attractive financing packages for their exports" (Brautigam, 2010, p. 7); however, a voluntary norm has been in place since 1978 to restrict this practice and since then financing modes such as export credits have not been able to be counted as foreign aid under OECD–DAC regulations (OECD, 1978).

As highlighted earlier, the Chinese Government utilises a range of financing mechanisms in its engagement with other developing countries. In addition, there is often a strong link between "aid" and "investment" projects, making accurate classification of a project difficult. The commercial aspect of Chinese aid is criticised as being self-interested; however, concepts such as the WTO's "aid for trade" and the EU's "policy coherence for development agenda" seek to address the links between aid, development and trade, as well as to encourage a coherent approach to international and bilateral relationships, and China's policies may be aligned in this regard. The possible appeal of China's approach of treating its relationship with a country more holistically needs to be explored further. Such an approach no doubt leads to concerns about corruption and lack of transparency in aid/investment deals, but arguably should not be disregarded too rapidly. As one PNG Government representative commented: "China's 'business approach' is appealing to governments – they know that China is also benefiting. It is more upfront" (Interview SP032, 8 October 2009).

The following examples from the resources sector in Fiji and PNG provide interesting insights into the relationship between Chinese investment and aid, revealing that whilst access to resources is an important part of China's engagement in the region, it is not a specific part of its aid policy and agreements. The Chinese-funded Nadarivatu hydrodam in Fiji was initially believed to be part of the Soft Loan Facility and reported as such by the 2009 Lowy Institute Report (Hanson, 2009), but my investigations reveal that this is not the case. The Fijian Government still maintains it is a concessional loan, but the Chinese Embassy in Suva, Sinohydro (the contracted Chinese company), and Fiji Electricity Authority (FEA), who took out the loan, all insist it is commercial. The fact that the loan is taken out with the China Development Bank (CDB) provides further evidence that this is the case, as CDB does not provide concessional aid funding. The construction of the dam is reported to cost a total of US$155 million, and while this is not an aid project, the Chinese company received support from the Chinese Embassy in Fiji for its application for the contract, for the establishment of crucial links with the Fijian Prime Minister's Office, and in forms such as donations of equipment to the local community. Sinohydro representatives acknowledge these advantages of being a SOE but suggest that "this is okay because China needs to support its companies because it is hard to compete with Western companies in the international arena" (Interview SP019, 27 September 2009).

There is also a new desire on the part of Fiji to capitalise on China's willingness to invest in the mining sector, with the chairman of Fiji's Mining and Quarrying Council predicting that mining will be the country's number one export earner within five years. A recent Radio Australia (2011a) program reported that the first set of native title lands has been leased to a Chinese company, Xinfa Aurum, which has begun mining the land and exporting bauxite to China to be used in Chinese aluminium products. This is a recent development that should be followed closely, particularly to see whether any Chinese foreign aid is used in support of the commercial activities.

Papua New Guinea is said to be on the edge of a major resources boom, with numerous mining projects and the mammoth LNG project likely to double the size of PNG's economy over the next decade (Morris, 2011). The largest Chinese investment is the US$1.4 billion Ramu nickel and cobalt mine project in Madang Province. Here, too, we see the Chinese aid and investment relationship at play. This is a *commercial* contract signed at the very top political level, undertaken by a Chinese state-owned enterprise, and supported – to a certain extent – by the Chinese Embassy in Port Moresby; it includes a key road in the mine area being funded through a Chinese Government aid grant[10] and reports of villagers from "mine affected communities" receiving scholarships and training opportunities under Chinese Government aid grants (Interview SP041, 13 October 2009). The project itself has been fraught with challenges (Smith, 2011), which complicates not only the relationship between the Chinese and PNG Governments, but also between the Chinese state and the commercial investor, one of its state-owned enterprises.

These brief examples indicate where a significant difference exists between Chinese foreign aid and Western aid – in the willingness for the Chinese Government to use its aid program to support Chinese commercial projects, either to help facilitate the agreement in the first place, or to assist when the company faces local community backlash, or more generally to support the investment itself. Importantly, this is *not* the same as the "aid and investment packages" or "resource-for-aid" deals that have featured in some other developing country regions. The types of support seen in the South Pacific are less formalised. Whilst there are still many instances of Western aid being "tied" or boomeranging back to the country's own contractors and Chinese aid mechanisms (at least the concessional loans) in fact require a certain amount of "tying" (it has to be win-win after all), the utilisation of an aid program to directly support a commercial investment is not (or should not be) part of DAC aid, and thus is a noteworthy distinction.

The benefits of this "aid and investment" link to China and (potentially) developing countries have already been outlined, but this dynamic can also present challenges for the Chinese aid program and its overall engagement in a country. Chinese companies contracted for aid (and investment) projects focus on getting them completed "on time and on budget". They do not see themselves as accountable for any "side effects" associated with their approach or tactics. This reflects the challenges of China's current system of overseas aid and investment projects. We are seeing an increasing disjuncture between a SOE's imperative on individual projects and the Chinese Government's broader strategic relationship with and goals in a country. In other words, the actions of Chinese companies are frequently negatively affecting the way China is perceived. In the eyes of local communities there is no distinction between Chinese Government "aid" projects and commercial investments; they are seen as the same thing because companies are "government supported" (Interview SP046, 16 October 2009). In the long run the approach and use of Chinese companies may be detrimental and may prompt China to develop different mechanisms for undertaking large aid projects.

Conclusion

This article has provided an analysis of China's foreign aid program and practices, particularly examining the objectives and mechanisms and the ways in which they are related to commercial interests and resource acquisition. In focusing on these dynamics in the South Pacific region, this paper highlights interesting aspects that serve to complement existing analyses of China's involvement in other parts of the world. Unlike in other regions, in the South Pacific there are (as yet) no examples of resource deals being explicitly part of Chinese aid. Current policies suggest that the resources sector is not an overriding focus or objective of Chinese aid in the region. A notable feature of Chinese aid in the region, however, is the way that the foreign aid program is being utilised to support commercial activities, including in the resources sector, particularly when problems have arisen. The nexus between aid and investment and commercial and state interests that features quite prominently in Chinese development assistance presents an interesting challenge for traditional donors who (in theory at least) separate "aid" from other forms of engagement with developing countries. It also, however, presents increasing difficulties for the Chinese state in managing the often-conflicting objectives, desires and practices of its state-owned enterprises and diplomatic apparatus.

LIST OF INTERVIEWS CITED

Code Number	Date	Location	Type of Stakeholder
SP004	11 September 2009	Suva, Fiji	Regional organisation
SP010	17 September 2009	Suva, Fiji	Traditional donor (bilateral)
SP011	17 September 2009	Suva, Fiji	Chinese official
SP013	18 September 2009	Suva, Fiji	Government body
SP016	24 September 2009	Suva, Fiji	Government official
SP017	24 September 2009	Suva, Fiji	Traditional donor (multilateral)
SP019	27 September 2009	Lautoka, Fiji	Chinese company
SP020	28 September 2009	Suva, Fiji	Government body
SP021	28 September 2009	Suva, Fiji	Embassy official
SP022	29 September 2009	Suva, Fiji	Civil society
SP026	5 October 2009	Port Moresby, PNG	Government official
SP027	6 October 2009	Port Moresby, PNG	Government official
SP030	7 October 2009	Port Moresby, PNG	Government official
SP032	8 October 2009	Port Moresby, PNG	Government official
SP041	13 October 2009	Port Moresby, PNG	Academic, UPNG
SP044	15 October 2009	Madang, PNG	Provincial government official
SP046	16 October 2009	Madang, PNG	Civil society
SP049	20 October 2009	Port Moresby, PNG	Traditional donor (multilateral)
SP052	22 October 2009	Port Moresby, PNG	Traditional donor (multilateral)
US057	19 April 2010	Washington, DC, USA	Government research officer
CN064	6 May 2011	Beijing, China	Traditional donor (bilateral)

Acknowledgments

This article was first prepared as a paper for the 'International Politics of Resources: China, Japan and Korea's Demand for Energy, Minerals and Food' workshop at the China Research Centre, UTS, 28–29 July 2011, and benefited from feedback from workshop participants, particularly Kate Barclay and Graeme Smith.

Notes

1. The strategy and its related initiatives include the promulgation of guidelines on outward FDI by countries and sectors, information regarding foreign countries' investment environment and

opportunities, delegation of authority by the central government to certain provinces and municipalities and further relaxation of foreign exchange controls on outward investment (Rutherford et al., 2008, p. 1).

2. For example, the strong involvement of the China Development Bank (CDB) in financing US$50 billion worth of large, long-term oil supply agreements with Russia, Kazakhstan, Brazil, Venezuela and Ecuador.
3. According to the White Paper, "complete projects" refer to productive or civil projects constructed in recipient countries with the help of financial resources provided by China as grants or interest-free loans. The Chinese side is responsible for the whole or part of the process, from study, survey, to design and construction, provides all or part of the equipment and building materials, and sends engineers and technical personnel to organise and guide the construction, installation and trial production of these projects. After a project is completed, China hands it over to the recipient country.
4. In contrast to the situation of many "traditional donors", there are no real domestic civil society pressures on the Chinese Government to give aid to other countries. The Chinese Government has, however, started to educate the Chinese public about its overseas aid program. Given the significant social and economic challenges still facing much of the population, this is partly to respond to negative public reactions to its aid program (see Brant, 2011a for more details).
5. Since the election of President Ma Ying-Jeou in Taiwan in 2008 there has been an informal agreement to not actively contest for each other's diplomatic allies.
6. Summary available from www.gov.cn/english/2006-04-06/content_246674.htm
7. A new agreement will reportedly be negotiated in late 2012.
8. As at November 2011 this loan had yet to be disbursed as the Chinese contractor was still awaiting release of PNG Government counterpart funds.
9. Confidential documents sighted by the author. These loan terms have now been extended by 10–15 years.
10. *Papua New Guinea Annual Budget Report 2011*, Volume 3, Section A, National Departments, Public Investment Project Number 2794: Usino Junction-Yamagi (Ramu), pp. 471–72.

References

Alden, Chris and Christopher R. Hughes (2009) Harmony and discord in China's Africa strategy: Some implications for foreign policy. *The China Quarterly* 199, pp. 563–84.

Brant, Philippa (2011a) China's aid transparency woes. *whydev.org* [Web log post]. Available at http://www.whydev.org/chinas-aid-transparency-woes/, accessed 21 March 2013.

Brant, Philippa (2011b) China releases first White Paper on aid. *Interpreting the Aid Review.* Lowy Institute for International Policy. Available at http://aidreview.lowyinterpreter.org/post/China-releases-first-White-Paper-on-Aid.aspx, accessed 21 March 2013.

Brautigam, Deborah (2009) *The dragon's gift: The real story of China in Africa* (Oxford: Oxford University Press).

Brautigam, Deborah (2010) *China, Africa and the international aid architecture.* African Development Bank Group Working Paper Series No. 107, April.

Callick, Rohan (2009) China and Taiwan end war over Pacific aid. *The Australian*, 10 August.

Chen, Deming (2008) Remarks at the investment, trade and tourism ministerial conference of the China–Pacific Island Countries Economic Development and Cooperation Forum. Available at http://english.mofcom.gov.cn/aarticle/translatorsgarden/famousspeech/200810/20081005818753.html, accessed 23 November 2010.

Commonwealth of Australia (2009) *Economic and security challenges facing Papua New Guinea and the island states of the Southwest Pacific.* Foreign Affairs, Defence and Trade References Committee Senate Inquiry, Official Committee Hansard, 19 June (Canberra: Commonwealth of Australia).

Crocombe, Ron (2007) *Asia in the Pacific Islands: Replacing the West* (Suva: IPS Publications University of the South Pacific).

Ding, Sheng (2008) To build a "harmonious world": China's soft power wielding in the global South. *Journal of Chinese Political Science* 3(2), pp. 193–214.

Dobler, Gregor (2009) Chinese shops and the formation of a Chinese expatriate community in Namibia. *The China Quarterly* 199, pp. 707–27.

Foster, Vivien, William Butterfield, Chuan Chen and Nataliya Pushak (2008) *Building bridges: China's growing role as infrastructure financier for Sub-Saharan Africa,* July (Washington, DC: The World Bank).

Gill, Bates and James Reilly (2007) The tenuous hold of China in Africa. *The Washington Quarterly* 30(3), pp. 37–52.

Greig, Helen (2010) China loan unlikely. *Cook Islands News*, 10 March. Available at http://www.cinews.co.ck, accessed 21 March 2013.

Han, Zhiqiang (Chinese Ambassador to Fiji) (2011) Lecture remarks, University of the South Pacific, Fiji. Available at http://www.gov.cn/misc/2011-05/05/content_1857860.htm, accessed 21 March 2013.

Hanson, Fergus (2009) *China: Stumbling through the Pacific* (Policy Brief: Lowy Institute for International Policy).

Hanson, Fergus and Jenny Hayward-Jones (2009) China's help may harm Fiji. *The Australian*, 23 April. Available at http://www.theaustralian.com.au/news/opinion/chinas-help-may-harm-fiji/story-e6frg6zo-1225701907555, accessed 21 March 2013.

Hanson, Fergus and Mary Fifita (2011) China in the Pacific: The new banker in town. Policy Brief, Lowy Institute for International Policy, April.

Herberg, Mikkal (2011) US perceptions of energy security and China's energy resource diplomacy. Paper presented at Conference on 'Resources Diplomacy Under Hegemony', co-hosted by University of Macau, The Hong Kong University of Science and Technology, and University of Alberta, Hong Kong and Macau, 20–22 January.

Jakobson, Linda (2011) New foreign policy actors in China [Audio podcast]. *Wednesday Lunch at Lowy.* Lowy Institute for International Policy. Available at http://www.lowyinstitute.org/Publication.asp?pid=1612, accessed 18 June 2011.

Jansson, Johanna (2011) *The Sicomines Agreement: Change and continuity in the Democratic Republic of Congo's international relations*. Occasional Paper No. 97 (Johannesburg, South Africa: South African Institute of International Affairs).

Kobayashi, Takaaki (2008) *Evolution of China's aid policy.* Working Paper No. 27, JBIC Institute.

Lancaster, Carol (2007) *Foreign aid: Diplomacy, development, domestic politics* (Chicago: University of Chicago Press).

Lesa, Mata'afa Keni (2009) White elephant and missed opportunity. *Samoa Observer*, 14 July. Available at http://www.samoaobserver.ws/index.php?option=com_content & view=article & id=10533%3Awhite-elephant & Itemid=104, accessed 20 March 2013.

Li, Xiaoyun (2008) China's foreign aid to Africa. Paper presented at workshop on 'Managing Aid Effectively: Lessons for China?', International Poverty Reduction Centre in China, Beijing, 27–28 March.

Lum, Thomas, Hannah Fischer, Julissa Gomez-Granger and Anne Leland (2009) China's foreign aid activities in Africa, Latin America and Southeast Asia. Congressional Research Service. Available at http://www.fas.org/sgp/crs/row/R40361.pdf, accessed 21 March 2013.

Maclellan, Nic (2011) Hot potatoes for Pacific trade policy [Web log post]. *The Interpreter*, 16 May. Available at http://www.lowyinterpreter.org, accessed 21 March 2013.

McCully, Murray (2011) New Zealand, Australia and China's rise. Opening address to public symposium 'New Zealand, Australia and China's Rise', Wellington, 6 April. Available at http://www.scoop.co.nz/stories/PA1104/S00126/mccully-speech-new-zealand-australia-and-chinas-rise.htm, accessed 10 April 2011.

Ministry of Commerce (2007) The situation of China's foreign aid in 2006. Department of Aid to Foreign Countries, People's Republic of China, 15 January. Available at http://yws.mofcom.gov.cn/aarticle/b/d/200701/20070104267249.html, accessed 21 February 2010.

Morgan, Wesley (2012) PACER–Plus: Where to now for regional trade policy in the Pacific? [Web log post]. *Development Policy*, 25 January. Available at http://devpolicy.org/pacer-plus-where-to-now-for-regional-trade-policy-in-the-pacific/, accessed 21 March 2013.

Morris, Matthew (2011) How can PNG fight the resource curse? [Web log post]. *Development Policy*, 27 May. Available at http://devpolicy.org/how-can-png-fight-the-resource-curse/, accessed 21 March 2013.

Naím, Moisés (2007) Rogue aid. *Foreign Policy* 159, pp. 95–96.

OECD (1978) *Recommendations of terms and conditions of aid* (Paris: OECD–DAC).

People's Republic of China (2003) *Chinese leaders on Sino-African relations* (Beijing: China Internet Information Center). Available at http://www.china.org.cn/english/features/China-Africa/82054.htm, accessed 21 March 2013.

People's Republic of China (2011) *China's foreign aid: White Paper* (Beijing: Information Office of the State Council). Available at http://www.scio.gov.cn/zxbd/wz/201104/t896900.htm, accessed 26 April 2011.

Porter, Edgar A. and Terence Wesley-Smith (2010) Oceania matters, in Terence Wesley-Smith and Edgar A. Porter (eds), *China in Oceania: Reshaping the Pacific?* (New York: Berghahn Books).

Quinn, Andrew (2011) Clinton says China seeks to outflank Exxon in Papua New Guinea. *Reuters*, 2 March. Washington DC.

Radio Australia (2011a) Mining predicted to be Fiji's biggest future income earner. *Pacific Beat*, 11 January. Australian Broadcasting Commission. Available at http://www.radioaustralia.net.au/pacbeat/stories/201101/s3110384.htm, accessed 13 January 2011.

Radio Australia (2011b) China no longer a positive Pacific force: Analyst. *Pacific Beat*, 21 January. Australian Broadcasting Commission. Available at http://www.radioaustralia.net.au/international/2011-01-21/china-no-longer-a-positive-pacific-force-analyst/229392, accessed 23 April 2012.

Rutherford, Jeff, Kate Lazarus and Shawn Kelley (2008) *Rethinking investments in natural resources: China's emerging role in the Mekong region*. Policy Brief (Phnom Penh, Copenhagen and Winnipeg: Heinrich Böll Stiftung, WWF and International Institute for Sustainable Development).

Shie, Tamara Renee (2007) Rising Chinese influence in the South Pacific: Beijing's "Island Fever". *Asian Survey* 47(2), pp. 307–26.

Smith, Graeme (2011) Chinese interests in Pacific nations: Mining ventures in PNG. *East Asia Forum*. Available at http://www.eastasiaforum.org/2011/05/19/chinese-interests-in-pacific-nations-mining-ventures-in-png/, accessed 24 May 2011.

Smith, Graeme (2012) Chinese reactions to anti-Asian riots in the Pacific. *The Journal of Pacific History* 47 (1), pp. 93–109.

Strauss, Julia C. (2009) The past in the present: Historical and rhetorical lineages in China's relations with Africa. *The China Quarterly* 199, pp. 777–95.

The Economist (2011) Charity begins abroad: Big developing countries are shaking up the world of aid. *The Economist*, 13 August.

Tukuwasa, Epeli (2011) Government to construct 3 more roads. *radiofiji.com.fj*. Available at http://www.radiofiji.com.fj/fullstory.php?id=36582, accessed 10 May 2011.

Varrall, Merriden (2012) Chinese aid as a tool of soft power?, in Hongyi Lai and Yiyi Lu (eds), *China's soft power and international relations* (London: Routledge).

Vines, Alex (2011) Oil counts: The role Angola plays in current Sino-US relations. Paper presented at Conference on 'Resources Diplomacy Under Hegemony', co-hosted by University of Macau, The Hong Kong University of Science and Technology, and University of Alberta, Hong Kong and Macau, 20–22 January.

Wen Jiabao (2006) Win-win cooperation for common development [Keynote speech]. China-Pacific Island Countries Economic Development and Cooperation Forum, Nadi, Fiji, 5 April.

Wesley-Smith, Terence and Edgar A. Porter (2010) *China in Oceania: Reshaping the Pacific?* (New York: Berghahn Books).

Windybank, Susan (2005) The China syndrome. *Policy* 21(2), pp. 139–58.

Yang, Jian (2009) China in the South Pacific: Hegemon on the horizon? *The Pacific Review* 22(2), pp. 139–58.

Yang, Jian (2011) *The Pacific Islands in China's grand strategy: Small states, big games* (New York: Palgrave Macmillan).

Zhang, Yongjin (2007) *A regional power by default: China in the South Pacific* (Auckland: New Zealand Asia Institute).

Nupela Masta? Local and Expatriate Labour in a Chinese-Run Nickel Mine in Papua New Guinea[1]

GRAEME SMITH

The University of Sydney

Abstract: *Studies of mining projects in Papua New Guinea, since the development of the Panguna copper mine in Bougainville during the 1960s, have contributed to our understanding of the politics of interactions between resource companies, host governments and landowners. The Ramu Nickel mine, situated in northern Papua New Guinea, is China's largest investment in the Pacific to date at US$1.4 billion. The project is managed by a state-owned enterprise, China Metallurgical Corporation, and financed by China ExIm Bank. This venture presents an opportunity to understand Chinese resource investment in a comparative perspective. While many issues, such as conflict over land, internal migration, and the limited involvement of the Papua New Guinean state, are constant, one aspect specific to Chinese resource investment is the use (or non-use) of host country labour, and the high proportion of Chinese labour employed at the mine sites. This practice differs from the relatively limited, short-term use of expatriate labour common to Western mining projects in developing countries. The attitudes and experiences of local and Chinese workers and managers will be examined to determine what is new in this approach to resource extraction.*

From the perspective of many observers, Chinese resource investment in developing countries is different from Western resource investment in terms of its impact on local governance and local communities. It is often characterised in the popular press in terms of dubious business practices, interference in domestic political processes and "prison-like" working conditions for local workers, which are argued to stem from the attitudes of Chinese mine managers towards the populations of the host country (HRW,

2011). Other researchers argue that such reports are racist "contribution[s] to China bashing", while "in the main Chinese firms operate like other foreign investors in the industry" (Sautman and Yan, 2011, p. 10). Yet the working conditions of the Chinese workers in these spaces are rarely discussed, due to barriers of language and access. Very little is known about the relationship between ordinary Chinese workers and their local counterparts. The Chinese "management" is usually referred to as an undifferentiated whole, glossing over the complex politics that exist within any organisation, state-owned or otherwise.

This is not to argue that poor working conditions are not common in Chinese resource investments, but simply to note that only half of the story is being told, and that some of the most interesting aspects of the new wave of Chinese outbound direct investment (ODI) are being overlooked. One aspect that is specific to Chinese resource investment is the use (or non-use) of host country labour, and the high proportion of Chinese labour employed on site for long periods of time during the life of these resource projects. This is quite different from the short-term use of expatriate labour common to Western resource projects in developing countries which is generally limited to the construction phase when the local supply of skilled labour is insufficient (e. g. the PNG LNG project run by Exxon-Mobil, which required 18,000 workers during its construction).[2] Although much of the detail in this paper relates to the construction phase of the Ramu Nickel mine project, by 2012 there was no evidence of a significant shift towards a reduction in the proportion of Chinese labour employed on the project as this phase was coming to an end. A new township settlement to house a predominantly Chinese workforce recently opened just to the east of the Basamuk refinery plant, with the long-term aim of creating a new commercial hub for the Rai Coast region. Similarly, permanent housing for the workers and management has been built on-site at the company's headquarters in Madang, and at the Kurumbukare mine site, rather than relying on the fly-in fly-out (FIFO) model seen at many mine sites during the production phase. Aside from at the very senior levels of management (vice-presidential level and above), there is no evidence of a FIFO approach being adopted. In historical terms, a comparable example would be the extensive use of white Afrikaans labour alongside African labour in British-run South African gold mines during the nineteenth century (Lynch, 2004), with an analogous class divide between the Beijing-based management and the provincial mine workers, layered on top of a clear racial divide within the mining workforce.

Very little is known of how such new enclaves function. Do barriers of language and race break down over time, and is there any possibility of a shared class identity developing between local and expatriate workers, who experience similar working conditions? This paper will examine daily interactions between Chinese and local workers at the Ramu NiCo mine and refinery, a nickel laterite mine situated in Madang province, northern Papua New Guinea (PNG). Based on interview and survey data collected during the construction phase of the project, this article examines the attitudes and experiences of Papua New Guinean and Chinese workers and managers to flesh out the unexpected outcomes arising from this approach to resource development.

Detailed anthropologies of mining in PNG are difficult to produce, due to logistical problems in reaching the mine sites to collect data, and the reluctance of mining companies to allow independent researchers to spend long periods of time at the mine site, particularly given the poor environmental and social track records of many mining projects in the history of PNG. Nevertheless, the work of Benedict Imbun (2000) and Glenn Banks (1993) at the Porgera Gold Mine in Enga Province, Colin Filer (1990) at the now abandoned Panguna Copper Mine in the Bougainville Autonomous Region, and Martha MacIntyre and Simon Foale (2004) at the Lihir Gold Mine in New Ireland Province and the now-exhausted Misima Gold Mine in Milne Bay Province provide an excellent starting point to draw lessons on the unique nature of the mining industry in PNG, the world's most ethnically and linguistically diverse nation. This paper will draw inspiration from Chris Ballard and Glenn Banks' (2003) challenge to move beyond the mining company lurking monolithically and often menacingly in the background, and to pay closer attention to the internal structure and politics of the mining company, and what effects these dynamics have upon the Chinese and Papua New Guinean workforces at the Ramu mine.

Previous research on the impact of Chinese resource investment on local communities in PNG is limited, with the notable exception of Mike Wood's (1995) piece examining the construction of European and Chinese identities in the interaction between Malaysian Chinese loggers and local landowners in Western Province. In terms of the practices of Chinese-owned mining companies operating abroad, Ching-Kwan Lee (2009) has examined the impact of Chinese mine ownership on labour practices in Zambia, but there are few other studies on which to draw, partly because of the novelty of the phenomenon, and partly because of the difficulty in accessing Chinese mine sites.

This paper will draw upon surveys of Papua New Guinean workers at the Ramu mine site and refinery sites, conducted between December 2009 and July 2010 with the assistance of student enumerators from Divine Word University. Over 300 completed surveys were collected. This paper will examine the qualitative data set, obtained through open-ended questions, while some descriptive statistics will also be provided. It was not possible to conduct a formal survey of Chinese workers, but I was able to engage in unstructured interviews with mine workers at the Kurumbukare mine site, the Basamuk refinery site, the pipeline construction headquarters, and the company headquarters in Madang and Beijing between 2009 (during the construction phase) and 2012 (the beginning of the production phase). The blogs of mining workers, many of whom are prolific (there is little in the way of entertainment at the Ramu mine sites) also provide useful supplementary information.

The Peculiarities of Ramu

To date, the Ramu mine is the largest single ODI project undertaken by a Chinese company in the Pacific, with an estimated investment of US$1.4 billion, a projected output of 32,800 tonnes of nickel, and a mine life that is expected to last between 20 and 30 years.[3] The majority owner of the Ramu NiCo Project is China Metallurgical Corporation (MCC), which, together with three Chinese nickel consumers that act as "silent" partners (the "3 Js": Jinchuan Group, Jinlin Ji'en Nickel Industry Corporation, and Jiuquan Iron and Steel Group), owns 85 per cent of the venture. Most of the balance is owned by the original holder of the mining lease, Highlands Pacific (a Brisbane-based mineral exploration company listed on the PNG and Australian stock exchanges), with 8.56 per cent. Smaller stakes are held by the Mineral Resources Development

Company (MRDC, http://www.mrdc.com.pg), a PNG government agency which holds project equity in trust for other national stakeholders, in this instance for the local landowners and the local-level governments in the mine-affected area with 3.94 per cent; and the company Mineral Resources Madang Ltd (MRM) with 2.5 per cent, which holds project equity in trust for the local landowners (see Figure 1).

Several different corporations have held the exploration lease since the 1960s, and initially their interest was in neither nickel nor cobalt, but in chromium.[4] Chromium was in demand during the Cold War era of the 1960s and 1970s, having been identified by both the US Congress and the European Community as a "critical nonfuel mineral" (Buijs and Sievers, 2011, p. 3). [5] In a neat symmetry, following the global shift of manufacturing to China, the PRC State Council identified nickel and cobalt as key minerals, making it easier for Chinese companies to raise finance to develop such mines. With the rising price of nickel (largely due to the increasing demand for stainless steel, the final destination of 60 per cent of nickel ores), the gradual depletion of readily accessible nickel deposits in regions such as New Caledonia, and the development of high pressure acid leaching, nickel laterite deposits such as the one at Kurumbukare have become viable. The process of extracting nickel laterites is technically complex (Brand et al., 1997) and has confounded at least two Australian mining ventures in recent years, the Murrin Murrin and Ravensthorpe projects in Western Australia. In addition to the complexity of the refining process, the Ramu project faces the additional challenge of transporting the nickel ore 134 kilometres from the Kurumbukare mine site to the Basamuk refinery site via a pipeline, traversing rugged terrain. From there, the mixed nickel hydroxide intermediates will be transported to China for final processing to nickel and cobalt products at an unknown location. The nickel tailings will be "placed" via a pipe into Astrolabe Bay, despite legal proceedings by landowners from Basamuk, which delayed commissioning for nearly two years.

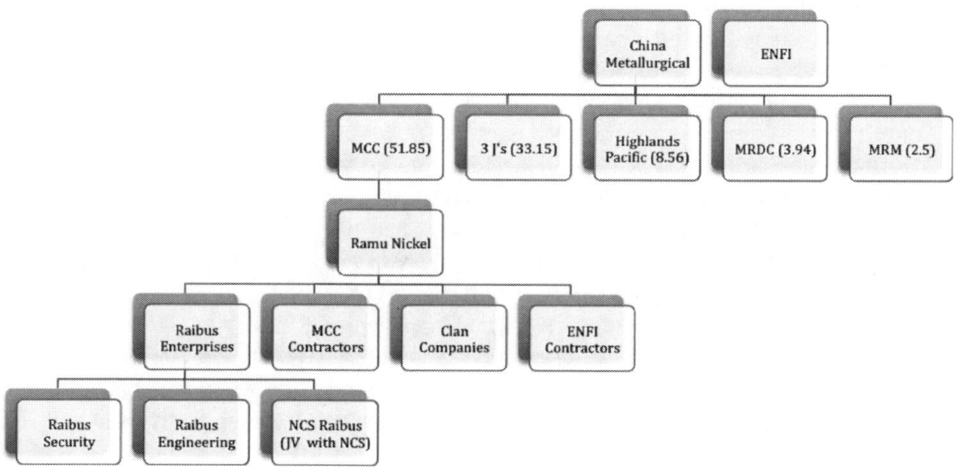

Figure 1.

In terms of literature examining China's ODI, the Ramu project is likely to be a candidate for the "infrastructure for resources" model of some Chinese resource investments, whereby large Chinese companies effectively bring the state with them. This will typically involve a package of loans for infrastructure, provided by China ExIm Bank or China Development Bank (CDB). Hardware such as roads, dams or pipelines will be built to support the resource project. Simply in terms of size, MCC seems to be large enough to "bring the state", but it has largely failed to do so. It is a "Fortune 500" company; it comes under the auspices of the State-Owned Assets Supervision and Administration Committee (SASAC), and with the encouragement of SASAC, recently listed on the Hong Kong and Shanghai stock exchanges. Yet due to its history and leadership, MCC does not have political clout comparable to CNOOC or other large oil companies. Its managing director, Shen Heting, complained that the level of government support MCC had received for its outbound investments was like "one hair of nine oxen" (*jiu niu yi mao* 九牛一毛) (Yan, 2010a). While large oil companies have more political muscle (and a higher formal rank) than many of the ministries which are meant to oversee them – as is seen in many "resources for infrastructure" projects driven by state-owned oil companies in Africa – MCC, as an offshoot of the Ministry of Metallurgy (*yejin bu* 冶金部), lacks the political influence to bring the state with it. To date, China ExIm Bank has been more active in PNG than China Development Bank,[6] supporting four large concessional loan projects as part of a $375 million soft loan facility to the Pacific as a whole, and providing finance for the Ramu Nickel project. None of the loan projects are related to the mine, and a proposed $5 billion infrastructure loan facility will provide little or no benefit to the Ramu mine (Radio Australia, 2012; Smith, 2012). Unlike the provincial and central governments, Ramu Nickel *has* upheld its side of the social and economic development bargain laid out in the Memorandum of Agreement (MOA) – which is still being negotiated 12 years after the initial agreement between Highlands Pacific, provincial and national governments, and landowner groups – despite the fact that MCC was not party to the initial MOA. The absence of large-scale infrastructure development associated with the project does, however, leave open the question of whether PNG is getting value for the concessions made in the 2005 Master Agreement (which built on the 2004 Framework Agreement, or *qujia xieyi*), signed by the Somare Government and MCC, which includes a 10-year tax holiday and a reduced rate of royalty payments. The current Prime Minister, Peter O'Neill, has admitted in a radio interview, "We have not done a good deal whereby we have not taken up more equity for our country" (*Post Courier*, 9 October 2012).

A further peculiarity of MCC is Shen Heting himself, who in addition to acting as the company president is also the party secretary and vice chairman of MCC's parent China Metallurgical Group Corp. In many ways, Shen is a product of the culture of the former Ministry of Metallurgy, which due to its heavy industry focus was one of the most privileged and loyal ministries during the Maoist era. Proximity to the state brought rewards, and this pervades the management culture, with management instructions still conveyed in yellow envelopes. Several authors have argued convincingly that Chinese resource companies pay lip service to the central state's concern for "resource security" as a way of extracting concessions, while in practice being driven largely by the profit motive (Shankleman, 2009; Downs, 2011). MCC's director, however, seems to believe that he is acting in the interests of the nation in securing resources, and thus he should be fully supported by the state, regardless of whether MCC's overseas ventures are profitable. In part, MCC's political marginalisation represents a cooling of enthusiasm for large-scale ODI ventures, fuelled by reports that up to two-thirds of Chinese investors abroad are loss-makers (Li et al., 2008).

As Shen explained in an interview with *New Century Weekly*, "Central government enterprises that secure mines overseas are in reality securing resources for China" (Yan, 2010b). In the same article, Shen explains MCC's approach to overseas resource investment – which involves a variety of minerals projects in a variety of regions[7] – in the following terms:

> High quality resources have already been taken by others, so at present, we can only go wherever there are still good resources. Which regions we invest in is not up to us to choose. As for risks, we can only rely on the support of the central government and our own efforts to avoid them. As long as it is a good resource, we will go for it (*zhi yao shi hao ziyuan, women dou na* 只要是好资源, 我们都拿).

In response, an analyst from Northwest Securities noted, "The feeling I get is that they [MCC] will sink their teeth into any mine they come across, without a clear understanding of what exactly they want to develop" (Yan, 2010b). By contrast, many large multinational mining companies (such as Xstrata, whose nickel division is located in Sydney, with copper operations run out of Brisbane) split their companies up on the basis of the metal being extracted. Part of this dissonance may arise from the fact that MCC is not, at core, a mining company. Domestically, its main focus has been construction and real estate. To an extent, it is learning about the mining industry from its experiences abroad.

Before examining the operations of the Ramu project at the ground level, another peculiarity is worth mentioning, as it provides an unexpected framework for this analysis. Due to the emergence of MCC from the Ministry of Metallurgy, the corporation operates through a large number of provincially based headquarters, which roughly correspond to the former provincial offices of the ministry. At the time of the survey, the Kurumbukare mine site was overwhelmingly staffed by contractors from MCC 19 and ENFI (Engineering Non-Ferrous Institute) 19, which were based in Panzhihua, a city in southern Sichuan province which is home to the Panzhihua Iron and Steel Group (Pangang), one of China's largest mines. The company recently relocated to the provincial capital Chengdu, but the majority of workers still hail from Panzhihua. These contracting companies are financially independent from the parent company, MCC, and are generally brought in to address a specific task (in the case of MCC 19 and ENFI 19, to construct the mine site). Due to the relative complexity of the work at the Basamuk refinery site, there is a more heterogeneous group of contractors engaged there, with contractors from Tangshan's MCC 22 and ENFI 22, Shanghai's MCC 20 and ENFI 20, and Anshan's (Liaoning Province) 3MCC (but at the time of writing, contractors from 3MCC had permanently returned to China).

This fragmentation of contractors means that each work site has a distinctive work culture, and a distinctive set of local identities, based on shared work histories, shared cultural preferences, and often a shared local dialect. While in its public announcements the Beijing-based parent company tries to give the impression that it is in complete control of the mining operation, the day-to-day operation is in the hands of the contractors, who are financially independent. This fragmented contracting system, whereby each contractor is independent and has no long-term involvement in the project, results in a short-term, limited liability approach to project management. It leads directly to pressuring local and Chinese workers to undertake longer hours to fulfil contracts ahead of time, and conversely to reduce expenditures on wages, safety equipment and accommodation.[8] As one provincial government official observed:[9]

> You know, landowners and the government, we play around, uh? Talking about this and that. But for them, they have no time to waste. They are here in PNG, and then get out. It's financial pressure. Because the finance is kicking their asses to hurry up and get it done quickly. They have to repay the loans. But many big projects are not given to Papua New Guineans, all given to China.

In a perfect regulatory regime, such pressures would not be a source of tension, but local and provincial governments in PNG have limited capacity to enforce their own regulations governing wages, workers' safety or land titling,[10] similar to the situation at other Chinese mining enclaves in Africa (Haglund, 2009, pp. 641–45). Other than police, whose incomes are generally supplemented, or paid in full, by the mining company, and a single health worker at the Basamuk clinic, the government has no presence at the mine site. These pressures are at the heart of the grievances of local workers, as we shall see in the following section.

Nupela Masta: Papua New Guinean and Chinese Perspectives on the Ramu NiCo Labour Regime

While the focus of analysis in this paper is primarily qualitative, a few initial descriptive statistics will provide background to the local workers' perspectives on working conditions at the Kurumbukare mine site and the Basamuk refinery site. The majority of disputes involving PNG workers at both sites revolve around levels of remuneration, and when asked whether they were satisfied with their level of payment, 90 per cent of workers at Kurumbukare (N = 145), and 87 per cent of workers at Basamuk (N = 156) stated that they were either "dissatisfied" or "very dissatisfied" with their salary.[11] The median fortnightly wage of workers interviewed at the Kurumbukare mine site was K260,[12] while those at Basamuk refinery site were paid K200.4, roughly equivalent to the minimum wage of K201.60 per fortnight (which was announced on 21 January 2010, just before the survey took place). Many workers were paid considerably less, with some reporting fortnightly salaries less than 100 kina. Nearly all workers stated that they had experienced some form of industrial action, and despite a negative experience with the setting up of a union at the mine site in 2008 whereby a man from the highlands province of Chimbu collected union dues from most workers at Kurumbukare and then promptly disappeared, close to nine in ten local workers expressed a desire to join a union if it were possible. Despite a litany of complaints about their working conditions, outlined below, it is striking that 64 per cent of workers at Kurumbukare and 70 per cent of workers at Basamuk indicated that they intended to continue working on the Ramu NiCo project. At the beginning of the production phase (late 2012), more than a thousand workers were employed by the project, nearly two-thirds of them Chinese. Many more local and Chinese workers were employed at the peak of the construction phase in 2009 (around 4,000 workers; reliable estimates are not available). The percentage of Chinese workers has remained fairly constant, at more than 60 per cent.

Previous researchers have found that a large source of tension for Chinese overseas enclaves was the communication barrier between local and Chinese workers (Lee, 2009, pp. 653–55) and our survey results bore this out strongly. Nearly every local respondent nominated this as the main difficulty in working with his or her Chinese co-workers and managers. One worker at Basamuk, who enjoyed good relationships with his Chinese co-workers, explained, "other Chinese labourers are good, we have fun, share smokes. However when it comes to working we are deaf and mute. We use signs instead of words. That is very challenging". The communication barrier was viewed as more of a problem in dealing with Chinese managers, and there was often an implication that the managers were "selectively deaf" in not understanding English or Tok-Pisin. In discussing the efforts of MCC to meet the PNG Government's language requirements by sending their Chinese workers to undertake English classes at the local university, a senior provincial government official posited:[13]

> They know, but they pretend. They know English, but they pretend they don't know English. And many of the MOA they signed with Department of Labour [and other parties], it's not working for them to take up classes at DWU [Divine Word University], they are not honouring those agreements. But we, in the government, because the national government has made the commitment, we are just trying to do our best.

The communication barrier added to the tension of negotiating over pay and safety issues, with many workers saying that communication with their managers became worse when under stress. This also spread to other issues. As one worker explained:

> Most Chinese workers emphasise more on work. There is not enough time for rest. So I am not in good terms with them. Poor understanding and communication widens. The Chinese workers and bosses do not observe the national public holidays and so it is difficult to work as Papua New Guineans should observe those days by not working.

Attitudes towards time were a particular source of tension, with many Chinese workers seeing themselves as the bringers of modernity to a workforce that had yet to experience the benefits of China's "reform and opening" (cf. Lee, 2009, pp. 654–59). This theme was developed by the then Prime Minister of PNG, Michael Somare, in an interview with Al Jazeera, in answering criticism about why migration and labour law requirements had been waived for large numbers of Chinese labourers during the construction phase of the project:[14]

> They thought the government was making a mistake by just employing Chinese. Because to get a new industry going, you know, Chinese are workaholics; they

> work from six o'clock in the morning, even at night, until they complete the task. Ask the Melanesians. We work when time permits us. When we feel that after ten o'clock when we are tired, we go and sit under a tree. Our work ethic has to be changed in this country; we have to work much harder.

A Chinese worker at the pipeline camp at Naru believed that the gap in attitudes towards work was unlikely to ever be bridged, due to the different historical experiences of Chinese and Papua New Guinean labourers:[15]

> They tend to be overly dependent on others. I think it comes down to their traditional way of life. It's rich in resources here. They can say, "I don't need to work, I still have bananas, pineapple, coconuts…" Even though their living conditions aren't great, they have no concerns about subsistence (*wenbao* 温饱). They've never experienced droughts or wars; they don't worry about whether there will be food to eat tomorrow. They don't know scarcity. They are used to abundant natural resources; they depend on them. So, when the Chinese arrived, they were unsettled, and the relationship is tense.

Paradoxically, while no one questions the work ethic of Chinese miners, observers in the mining industry expressed the opinion that the project was far more labour intensive when compared to other projects.

In one crucial sense, the Chinese enclaves in PNG are different from those in Africa,[16] because China lacks a substantial history of engagement with PNG, unlike in countries such as Tanzania or Zambia, where China has a long and generally positive history of engagement (Alden et al., 2008; Brautigam, 2010). On the Chinese side, this leads to an intriguing disconnection, as the enclave workers, many of whom are hired by labour contracting companies, arrive in PNG with no preconception of their destination. Indeed, most of them joke that before they arrived, they thought New Guinea was in Africa, somewhere close to Guinea. Whereas "precolonial" migrants to the Pacific from Europe in the nineteenth and twentieth centuries could draw on missionary accounts or the literature of Robert Louis Stevenson to provide a framework for their experiences before they arrived in a colony (Steinmetz, 2007), a search of *Chinese Academic Journals Online* (the most comprehensive database of Chinese language academic literature) reveals fewer than ten articles about PNG.

Many Chinese workers sought to understand their experiences in PNG through the lens of China's own frontier regions, where many infrastructure and resource development projects are situated. In May 2009, the Ramu mine experienced two riots at its Basamuk refinery plant after a PNG national worker was thrown from a trailer towed behind a tractor, resulting in horrific injuries to his lower body. The failure to use a helicopter to evacuate the worker, who was the nephew of a prominent landowner, sparked the initial fight, and when a handful of Chinese workers attacked the police station for not doing enough to protect them a riot erupted. There were injuries to more than 20 Chinese workers, with several treated at Modillon Hospital in Madang. In discussing this incident, and the (unrelated) nationwide anti-Asian riots that followed it, in interviews, many workers asserted that PNG was "just like Xinjiang", which had seen bloody ethnic unrest earlier in 2009. They also often likened the constraints on their daily lives to those faced in ethnic Tibetan regions, where some had previously worked, and where they also faced the challenges of different local cultures and work practices, and a local population that they were physically and linguistically isolated from (Hansen, 2005). Much like mining camps in China's frontier regions, they were provided with, and were in the process of creating, an environment that guaranteed that their experience of Papua New Guinean culture would always be limited. They ate Chinese food in a Chinese-only mess, called their families every other day, watched Chinese DVDs, spent time on social networking sites such as QQ, and gambled for small stakes with Chinese currency. Such separation is dictated by concerns for the workers' safety. In larger Chinese enterprises, regulations prohibiting mixing with the local population are usually enforced regardless of whether the host environment is dangerous or not. In Samoa, one of the safest countries in the world, a worker living in a Chinese construction camp explained, "We'd love to go out at night. It's really boring here. But our company has operations in many countries – Syria, Pakistan, Vietnam – and the regulations are enforced uniformly, regardless of where you are posted".[17]

The related issues of land claims and migration can be fraught for mining projects, and the five-decade history of expectation surrounding the Ramu project and the nearby Yandera copper-molybdenum project means that land ownership is highly contested, especially in Kurumbukare (Zimmer-Tamakoshi, 1997). It presents a classic case of a relatively marginalised ethnic group, the Gende people, poised to benefit from a shift in their relative regional power, as has been seen for the Ipili people at Porgera (Banks, 2008, p. 28). Nevertheless, the management of Ramu NiCo did not take the issue seriously, and insisted that landowners take their complaints to the central government, well after it became apparent that the PNG state was largely absent from any of the project areas. The prospect of "landowner benefits" has led to migration into the area from elsewhere in PNG (Wabag, the capital of Enga Province in the Highlands, is frequently mentioned) in order to take advantage of the failure of the government to undertake land titling. In November 2011, Land Title Commission hearings to resolve disputed claims finally took place in Madang. As has happened elsewhere in PNG, this has led to acts of violence against outsiders, usually disparagingly referred to as "settlers", largely to the bemusement of Chinese management and workers.

Disputes between landowners, mine management and government over local employment are common in PNG, and often have major consequences for mining ventures (Filer, 1990; Imbun, 2000; Ballard and Banks, 2003; Banks, 2008; Bainton, 2009). Surveys reveal that this tension is present in the workforces at both Kurumbukare and Basamuk. One respondent from Basamuk echoed a theme common to all mines in PNG, saying, "I do not like workers from other areas working here, they are occupying the space for the local landowners to work in the mine. The priority should be for the locals and not outsiders". Nevertheless, the majority of sentiment is positive, with most respondents referring to workers from elsewhere as their "brothers and sisters". It seems that the Chinese management and workers are filling the role of "outsiders" that elsewhere leads to inter-ethnic tension between different groups of Papua New Guineans, at least within the mining enclaves.

While there has been a history of Chinese migration to PNG dating back to the nineteenth century (Wu, 1982), the Chinese workers remain "dislocated from local narrative and practices" and "somewhat alien", exactly as Mike Wood found in his study of local perceptions of Malaysian Chinese logging workers in Western Province (Wood, 1995). Just as Wood cited Fanon's famous maxim, "You come too late, much too late, there will always be a world – a white world between you and us", local workers constantly compare the Chinese to previous colonisers and investors from the West. The comparison is often unfavourable, and leaves the sense that a fictitious history of positive Western colonisation and civilisation is being created to make sense of negative interactions with present-day Chinese investors, and the failings of the Papua New Guinean state. The previous Australian colonial regime, which handed over power in 1975, placed restrictions on the activities of Asian residents in PNG, entirely consistent with the White Australia Policy, which was only abandoned in Australia in 1972. Some Western expatriates fuel such negative comparisons. Less positive media coverage is given to Asian ODI by News Limited's *Post Courier* than by the Malaysian-Chinese owned *National* (Tumu, 2007). Such a comparison leaves many Chinese managers bemused. They disavow any colonising or civilising mission, often stating, "We just want the nickel". Initial interactions between the company and white expatriates were often awkward, as one (white) PNG national who works with the company related:[18]

> My first experience with the Chinese was pretty funny. We had our first meeting and all the directors [of MCC] were there. They expected to see all PNG Nationals. I'm a citizen, and when I walked in, Mr X was sitting directly opposite me at the table, and he had this inscrutable Chinese face on, and these jet black eyes staring right at me. I just played it very cool and didn't say anything. They nearly fell off their chairs when I walked in. Because one of the Chinese after that said to me, under no circumstances will we be employing any Australians. They had a real thing about Australians.

Reservations about hiring non-Chinese expatriates seem to remain. At the time of writing, less than 1 per cent of the project workforce is expatriates of European origin. As long as the language of communication within the management is Chinese, which is unlikely to change (at the time of writing, the effective director of the Basamuk site is an engineer with a limited grasp of English), there is a barrier to either expatriate or local staff being effective in higher levels of management. This also limits the scope for Ramu Nickel to integrate with or learn from other mining projects in PNG. As one observer noted, changing management practices is difficult if "you don't know what you don't know". Reinforcing this barrier, some expatriates employed by other mining companies in PNG were overly willing to dismiss the Ramu Nickel project.

This is not to suggest that the differences between Western mining companies' practices and those observed at Ramu NiCo are imaginary cultural artefacts. Our surveys reveal a disturbing array of health and safety issues, with common complaints being entirely preventable diseases such as dysentery and typhoid, and equally avoidable industrial accidents, such as eye injuries from welding and electrical accidents. Instances of penny-pinching behaviour abound; those relating to the provision of uniforms, transport to the mine sites and failure to replace damaged safety equipment are sources of resentment. In part, the difficulty for the Chinese developer arises because the Papua New Guinean workforce is familiar with mining practices elsewhere in PNG, either through direct experience or through talking to experienced co-workers. These shortcomings, which can easily be explained by China's own status as a developing country, the poor practices within China's domestic mining industry, or the lack of mining expertise within MCC, are often seen in racial terms, as Western mining companies run all comparable mining projects in PNG. As a community affairs officer recalled:[19]

> Most of the time the young ones are trying to look as if they know what's going on but in fact they don't. One of my colleagues in land was talking about a drill rig being set up in a garden and the damage that it caused, and we've got to pay compensation. And the guy who was Chinese and looking after the money asked, "What is a rig, what is a drill rig?" And the national manager said, "Did I hear you correct? You asked 'what is a drill?' and 'what is a rig?'" "Yeah what is a drill rig?" And the national said, "If you don't know what is a drill, I think you should get the hell out of here! You're in this industry, you should know this". I mean PNG, it's not that big compared to China, and everyone knows about mining because the country's economy and everything has been boosted by the mining operations, Bougainville and Ok Tedi and Lihir and Porgera. Most of them coming from China haven't heard about mining and they come here and everyone on the street knows about mines. You see that with health and safety. X has shut down quite a few posts, and the dorm cabins, which were not safe. They were leaving their toilets blocked. They don't have the safety standards required. In PNG people have come to expect the mining operations reach a certain standard of benefits and conditions, and these guys, zero, they haven't even come close yet.

Those mine workers with previous experience at other mines in PNG were asked, "How do the living and working conditions at Ramu NiCo compare to your previous employment with other mining companies?" Ten different mine sites were mentioned (Ok Tedi was mentioned most frequently), and all responses reflected negatively on Ramu NiCo.

The management of Ramu NiCo offers several explanations for why it falls short of the working conditions found at other mines in PNG, arguing variously that conditions will improve once the production phase is under way and the contracting companies return to China. They also often point out that the mine is run along a "low-cost" model, which makes the exploitation of a relatively low-grade lateritic nickel ore body viable. They are correct in pointing out that comparison with highly profitable gold and copper mines is unfair. Many local workers accept the logic that the construction phase is necessarily chaotic, not least because many of them have observed the incentives faced by the contracting companies. One former employee described a stand-off between MCC management and ENFI contractors in 2007, which nearly led to a fight outside the main bus stop in Madang:[20]

> We didn't quite get to the stage of punching up. We were about to and we just backed off. The contractors ENFI PNG were going ahead and doing work without consulting us, without us going to talk to the locals to get them aware that this [pipeline] will be coming. We have to compensate them if we do any damage on the land, especially with the clearing, and we have to record any damage and

> pay according to fairly generous rates. And sometimes, the contractors never told us, and just went in and did the clearing. By the time we got notice and got there, we were late, and we couldn't really know what was there, so we were obliged to pay what the locals were demanding, because we didn't have our records. That's what the primary reason [for the fight] was. So it got [to] the stage where it boiled up and our boss said, hey, you're the one going around causing trouble and making me look bad.

Chinese workers were also critical of the practices of some of the contractors engaged at the Ramu mine. One interviewee stated that the basic problem was that MCC was "not a mining company, it's a construction company, so they don't think long term". He identified different work cultures at the different sites, and like many respondents, linked the success or failure with the degree to which the companies had reformed their old socialist practices (cf. Lee, 2009, pp. 654–56):[21]

> MCC and ENFI 19 are better, as they were the first to change its structure, so they have a more market-based approach to project management. MCC 22 [based in Tangshan, Hebei province] at the refinery site is completely different; it's an old enterprise. When I visited the site for an inspection, everyone wanted to tell me what to do. But as soon as I pointed out problems, no one wanted to know about them… Sichuan has always been a populous province, where the pressures of subsistence forced people to make an effort, otherwise you wouldn't have anything to eat. But the other areas are close to the capital. The closer you are to the emperor, the more likely you are to see him as the sky; it's the traditional way of thinking. They spend all their time fighting amongst themselves over spoils, but when it comes to external matters, they ignore issues they should manage, and stick their nose in when it's not needed … thousands of years of official bickering hasn't changed; thousands of years of the feudal society, up to present-day socialism, officials haven't changed.

In a very different way, one of the more colourful landowner leaders from the Kurumbukare site came to similar conclusions about how the management treated the landowner umbrella group company, Raibus Enterprises, which was meant to look after spin-off business from the project:[22]

> It was mentioned in the MOA [Memorandum of Agreement] that all [spin-off] contracts should be given to the landowners. But in the beginning the Chinese came here, they had the document signed in Beijing with Prime Minister of the country, Michael Somare [the Master Agreement, signed in 2005]. And the Chinese came in and they're raised in a society, which is totally different from ours. They don't have landowners. They don't know what is a landowner. They signed this [Master] Agreement and they thought that the Prime Minister [Somare] is the Godfather of everything. They said go in there and just do your business. They came here and said, there are no landowners, the Prime Minister is the landowner. Land belongs to Somare. The Chinese do not recognise the landowner. They thought the Prime Minister is like their Prime Minister, I don't

> know, Mao Tse Tung or whatever! He gives an order and everybody dies for this order. We really had a hard time.

The perception that the Chinese company was overly close to the central government leadership was widespread among PNG nationals, and many interviewees confirmed that when there were conflicts with landowners or workers, Chinese managers would tell them "*graun bilong Somare*" (the land belongs to former Prime Minister Somare). As one landowner in Basamuk joked, "We were stunned. I mean, Somare's from Sepik [Province]. He has lot of mosquitoes, a swamp, perhaps some crocodiles. But not real coast land like this".[23] This preference for a state-to-state approach fits neatly with the management culture of the Ministry of Metallurgy, and manifested itself in behaviour familiar to scholars of Chinese politics. One Papua New Guinean manager described his experience with the speech-making techniques of the management company:[24]

> Everything has to go back to Beijing. At first I was wondering, "What the hell is this?" Then I realised this is basically a government institution. Like somebody else is controlling it in Beijing. [When Shen Heting came from Beijing], the meeting, it was basically mobilising support. Big banners. I was the only Papua New Guinean among all the Chinese and I felt so embarrassed about it. They talked and talked. All of them shouted. And I came out of it not knowing anything but I assumed it was more like "we're now going to start a project, we've got to get behind this thing". I'm not used to it. It's like rugby league team, you know. They psych you up before you get to the field. And Madame Luo [then president of Ramu NiCo] has got a lot of that too. During our monthly meetings, and if something is wrong, oh man. I was sitting there and just by the body language, you know something is wrong.

One interviewee wryly noted that Madame Luo seemed to believe that "there was no problem that a four-hour motivational speech couldn't cure".

A Shared Class Identity?

This eclectic mixture of traditional socialist management practices, combined with the high-pressure approach of the contracting companies, often left Chinese and Papua New Guinean employees equally bewildered. There are indications from the surveys of genuine friendships being formed between the two workforces, based on shared opposition to management practices, and, to an extent, on shared class identity. There is concern from the managers that the lower-ranking workers at the mine site, largely from rural or peri-urban areas of Sichuan province, were adopting many of the local customs, particularly the consumption of bush meat (snakes of the poisonous and fast-moving variety are popular) and the consumption of home brew, or *stim*. Bills posted by the management exhort workers to give up drinking outside festival periods, because this is leading to "social problems … conflict with locals, and endless complaints to police". Many Chinese workers have genuine affection for their (educated) PNG colleagues. One described the departure of a PNG colleague to work on another mine:

> It's already settled that he'll leave the company, all he has to do is sort out his termination, and in a couple of weeks we'll have to say goodbye. He's the best friend I've made here. Over the last two years, I've gotten to know his manner of speaking and his way of getting things done. He's helped me out with so many problems, and taught me so many things. But I fully support his decision to leave; I'm really happy for him.

Where Chinese and local workers are from a similar social class, and share similar levels of technical skills, there is evidence of meaningful relationships being formed. There are several reports from the surveys of local workers going out of their way, and even risking their lives, to help their Chinese co-workers in the event of accidents or violence at the mine site. One respondent recalled, "a Chinese expat almost got killed by naked wires running when he was working, he was in shock. I had to help him out, reported the matter, brought the victim to Madang General Hospital. We met our own medical expenses". While there have been serious injuries to PNG workers – such as the emasculation of a worker employed by an ENFI contractor at Basamuk refinery, which sparked a riot in May 2009 – to my knowledge, all deaths to date at the Ramu mine have involved Chinese workers. A variety of accidents have occurred, including falls, electrical shocks, misjudged river crossings, misdiagnosed malaria (after the worker returned to China) and at least one murder. While Chinese workers are better paid than their national colleagues, exploitation of workers cuts both ways.

It seems that there is a basis for anticipating that a new type of interaction, something approaching shared class identity, may eventually develop between local and Chinese workers at the Ramu mine site, or more likely, at the Basamuk refinery site. The length of tenure of PNG workers surveyed at the refinery site was longer than that of those at the mine site (nearly 20 months, compared to 13 months), and the settlement in the township at Basamuk is more permanent than that at the mine site. While turnover among younger Chinese technical personnel is high, many of the more experienced workers have been working in PNG for more than six years. Over time, as communication and cultural barriers are gradually worn down, a shared frustration with relatively low wages and the practices of the company's middle management is likely to build up. As one Chinese worker recalled:

> When we joined, we were taught that we were promoting the national interest. On the international stage, our future would be promising. When we heard these words, all of us swelled with pride. But after a few years of chaotic management and relentlessly stingy working conditions, this group chose to head for the door. With the Ramu project under internal and external pressure, the leaders busied themselves counting pennies, putting on a show, and sending "good news" back to MCC headquarters and to SASAC. As for whether young graduates stay, leave or progress, they don't care. But they'll be sure to claim penalties if you resign early.

It is striking that when asked why they want to continue working for Ramu NiCo, Chinese and local workers give the same answers: to support their families and pay for their children's (or child's) school fees and medical expenses. The life pressures faced by the two workforces are not dissimilar. They share a perception that despite the economic booms in their respective countries, driven by resources in PNG and manufacturing in China, their countries' elites are getting rich at their expense. As one Sichuanese worker remarked, while we were bumping along the road up to the mine site, "Our country is rich, but our people are poor".

Acknowledgments

The author would like to acknowledge the invaluable support of Sinclair Dinnen, Sonja Litz, Paul D'Arcy, Sue Rider, Kate Barclay, Patrick Matbob, David Hegarty, Heather Goodall, Hans Hendrischke and Colin Filer, the intrepid efforts of student enumerators from Divine Word University, in particular Eva Wangihama, Schola Chapok and Emil Yambel, and the guidance of two anonymous reviewers. All errors are of my own making.

Notes

1. "Nupela masta": literally "new rulers". The word "masta" was used to refer to the former white (variously German, British and Australian) colonial administrators of PNG.
2. In some regions, hybrid forms are emerging. In New Caledonia, where the local population is insufficient to staff the numerous mine sites, laws have been passed to allow Chinese construction workers at the Koniambo nickel mine to be employed under different working conditions from local and (French) expatriate labour. Workers of a further 30 nationalities (including Korean, Indian, Indonesian and Filipino workers) also operate outside the country's labour laws. Other mines in PNG have made extensive use of Asian (usually Filipino) labour during the construction phase. It is alleged that dual-track industrial conditions are also emerging in Australia, including on the Sino Iron project, where MCC is the lead contractor (Lucas, 2012).
3. For further details on the technical aspects of the project, see www.highlandspacific.com/pdf/Ramu_Nickel_Cobalt_Project.pdf, accessed 17 July 2011.
4. A Mining Development Contract was eventually signed between Highlands Pacific and PNG partners in 2000, resulting in the issuing of Special Mining Lease 8.
5. The EU's most recent list of "critical raw materials" includes antimony, beryllium, cobalt, fluorspar, gallium, germanium, indium, magnesium, niobium, platinum-group metals, rare earths, tantalum and tungsten (Buijs and Sievers, 2011, p. 8).
6. For details on CDB's involvement in Fiji, see Dornan, 2010.
7. In Afghanistan (Aynak Copper Mine, a $4 billion investment), Pakistan (Saindak Copper and Gold Mine; Duda Lead and Zinc Mine), Argentina (Sierra Grande Iron Mine), and Australia (lead contractor on the CITIC Pacific Sino Iron project).
8. A Human Rights Watch report notes similar problems with contractors at the Indian-owned Konkola Copper Mines in Zambia (HRW, 2011, pp. 67–68).
9. Author's interview, Madang, November 2009.
10. Local level governments (LLGs) have the power to enact by-laws relating to labour and employment, while provincial governments have some discretion to enact laws on land and land development issues. More power rests with the central government, and the Mineral Resource Authority's Chief Inspector of Mines, Mohan Singh, has applied significant pressure on the Ramu Nickel project to meet health and safety standards (Matbob, 2009).
11. Many Chinese workers were equally unimpressed with their salaries and working conditions. Existing employees were generally paid a 20 per cent loading on their base salaries, with some overtime. In contrast to other expatriate workers in the resource sector, who generally work two weeks on, one week off (in mining), or four weeks on four weeks off (in oil and gas), Chinese workers at Ramu Nickel are on a five months on, one month off roster (usually returning to China for their break). There are suggestions that this will improve when the project is running well.

12. At the time of writing, one PNG Kina was equivalent to 0.485 USD.
13. Author's interview, November 2009.
14. Al Jazeera International, 9 October 2009.
15. Author's interview, February 2010.
16. Other (smaller) enclaves have emerged at construction sites elsewhere in PNG, usually associated with infrastructure projects. The lifespan of these enclaves is much shorter, however.
17. Author's interview, Apia, Samoa, December 2011.
18. Author's interview, Madang, February 2010.
19. Author's interview, Madang, October 2009.
20. Author's interview, Madang, September 2009.
21. Author's interview, Madang, February 2010.
22. Author's interview, Madang, October 2009.
23. Author's interview, Madang, November 2011.
24. Author's interview, Madang, November 2009.

References

Alden, Chris, Daniel Large and Ricardo Soares de Oliveira, eds. (2008) *China returns to Africa: Rising power and a continent embrace* (New York: Columbia University Press).

Bainton, Nicholas (2009) Keeping the network out of view: Mining, distinctions and exclusion in Melanesia. *Oceania* 79(1), pp. 18–33.

Ballard, Chris and Glenn Banks (2003) Resource wars: The anthropology of mining. *Annual Review of Anthropology* 32, pp. 287–313.

Banks, Glenn (2008) Understanding "resource" conflicts in Papua New Guinea. *Asia Pacific Viewpoint* 49(1), pp. 23–34.

Brand, N.W., C.R.M. Butt and M. Elias (1997) Nickel laterites: Classification and features. *Journal of Australian Geology and Geophysics* 17(4), pp. , p. 81–88.

Brautigam, Deborah A. (2010) *The dragon's gift: The real story of China in Africa* (Oxford: Oxford University Press).

Buijs, Bram and Henrike Sievers (2011) *Critical thinking about critical minerals*. CIEP-BGR Briefing Paper. Available at http://www.clingendael.nl/ciep/publications/?id=8678 & & type=summary, accessed 28 January 2012.

Dobler, Gregor (2009) Chinese shops and the formation of a Chinese expatriate community in Namibia. *The China Quarterly* 199, pp. 707–27.

Dornan, Matthew (2010) Nadarivatu hydro scheme – A Chinese-Fijian partnership with pros and cons. *East Asia Forum*. Available at http://www.eastasiaforum.org/2010/05/22/nadarivatu-hydro-scheme-a-chinese-fijian-partnership-with-both-pros-and-cons, accessed 30 May 2012.

Downs, Erica (2011) *Inside China, Inc: China Development Bank's cross-border energy deals*. John L. Thornton China Center Monograph Series (Washington, DC: Brookings Institution). Available at http://www.brookings.edu/papers/2011/0321_china_energy_downs.aspx, accessed 28 January 2012.

Filer, Colin (1990) The Bougainville rebellion, the mining industry and the process of social disintegration in Papua New Guinea. *Canberra Anthropology* 13(1), pp. 1–39.

Haglund, Dan (2009) In it for the long term? Governance and learning among Chinese investors in Zambia's copper sector. *The China Quarterly* 199, pp. 627–46.

Hansen, Mette (2005) *Frontier people: Han settlers in minority areas of China* (Vancouver: UBC Press).

Human Rights Watch (2011) "You'll be fired if you refuse": Labor abuses in Zambia's Chinese state-owned copper mines. Available at http://www.hrw.org, accessed 6 November 2011.

Imbun, Benedict (2000) Mining workers or opportunist tribesmen? A tribal workforce in a Papua New Guinea mine. *Oceania* 71(2), pp. 129–49.

Lee, Ching-Kwan (2009) Raw encounters: Chinese managers, African workers and the politics of casualization in Africa's Chinese enclaves. *The China Quarterly* 199, pp. 647–66.

Li Jing, Guanghe Ran and Lijuan Wan (2008) Zhongguo qiye duiwai zhijie touzi jixiao bujia de yuanyin fenxi [An analysis of Chinese enterprises', lack of success in ODI]. *Shengchanli yanjiu [Productivity Research]* 3, pp. 69–71.

Lucas, Clay (2012) Foreign mine workers on half pay. *Sydney Morning Herald*. Available at http://www.smh.com.au/opinion/political-news/foreign-mine-workers-on-half-pay-20120528-1zffx.html, accessed 29 May 2012.

Lynch, Martin (2004) *Mining in world history* (London: Reaktion Books).

Macintyre, Martha and Simon Foale (2004) Politicized ecology: Local responses to mining in Papua New Guinea. *Oceania* 74(3), pp. 231–51.

Matbob, Patrick (2009) Ramu Nickel project blunders on. *Islands Business*. Available at http://www.islandsbusiness.com/islands_business/index_dynamic/containerNameToReplace=MiddleMiddle/focusModuleID=18846/overideSkinName=issueArticle-full.tpl, accessed 29 May 2012.

Post Courier (2012) PM lauds Ramu Nickel. Available at http://www.postcourier.com.pg/20121009/business01.htm, accessed 11 October 2012.

Radio Australia (2012) China offers help to rebuild PNG's Highlands Highway. Available at http://www.radioaustralia.net.au/international/radio/program/pacific-beat/china-offers-to-rebuild-pngs-highlands-highway/960820, accessed 23 July 2012.

Sautman, Barry and Hairong Yan (2011) Gilded outside, shoddy within: The Human Rights Watch Report on Chinese copper mining in Zambia. *The Asia-Pacific Journal* 9(52). Available at http://www.japanfocus.org/-Yan-Hairong/3668, accessed 28 January 2012.

Shankleman, Jill (2009) Going global: Chinese oil and mining companies and the governance of resource wealth (Washington, DC: Woodrow Wilson International Center for Scholars). Available at www.commdev.org/files/2580_file_DUSS_09323Shnkl_rpt0626.pdf, accessed 28 January 2012.

Smith, Graeme (2012) PNG: The six billion kina question. *The Interpreter*. Available at http://www.lowyinterpreter.org/post/2012/12/19/PNG-The-six-billion-kina-question.aspx, accessed 15 January 2013.

Steinmetz, George (2007) *The devil's handwriting: Precoloniality and the German colonial state in Qingdao, Samoa, and Southwest Africa* (Chicago: University of Chicago Press).

Tumu, Georgina (2007) *Representations of Asian investments in the PNG press*. Unpublished Honours thesis. Divine Word University, Papua New Guinea.

Wesley-Smith, Terence and Edgar A. Porter, eds. (2010) *China in Oceania: Reshaping the Pacific?* (New York: Berghahn Books).

Wood, Mike (1995) "White skins", "real people" and "Chinese" in some spatial transformations of the Western Province, PNG. *Oceania* 66(1), pp. 23–50.

Wu, David (1982) *The Chinese in Papua New Guinea: 1880–1980* (Hong Kong: The Chinese University Press).

Yan Jiangning (2010a) Zhongye haiwai zhao kuang lu [MCC seeks mineral wealth abroad]. *Xin shijie* 36, 4 September. Available at http://magazine.caixin.com/2010-09-04/100177070.html, accessed 15 January 2013.

Yan Jiangning (2010b) MCC chief defends overseas resource strategy. *Caixin* [New Century Weekly], 8 September. Available at http://english.caixin.com/2010-09-08/100178056.html, accessed 15 January 2013.

Zimmer-Tamakoshi, Laura (1997) When land has a price: Ancestral gerrymandering and the resolution of land conflicts at Kurumbukare. *Anthropological Forum* 7(4), pp. 649–66.

China, Natural Resources, Sovereignty and International Law

BEN SAUL

The University of Sydney

Abstract: *This article explores China's attitudes towards the regulation of key natural resources by international law, domestically and at the trans-boundary and international levels. It considers the impact of international law on China's own practices, and the contribution of China towards shaping international law. The article suggests that popular conceptions of a relatively isolated, sovereign absolutist China do not accord with contemporary legal realities, including in its dealings with natural resources. While China's construction of strong sovereignty shapes its attitudes towards legal regulation, practice also suggests that China adopts a nuanced approach which includes legal compromise, and a commitment to multilateral regulation or bilateral diplomatic settlement of issues previously within the competence of national governments. China is often an active and constructive participant in contemporary law-making, even if – like all countries – it also seeks to instrumentally use international law.*

Introduction

China's approach to international law is commonly critiqued for its preoccupation with defending a relatively absolutist understanding of sovereignty and the rights flowing from it. China's attitude is seen as a barrier to both the penetration of international norms within China and transnational cooperation on shared problems. China is thus constructed as a recalcitrant: resistant to human rights and democracy; cautious about humanitarian intervention, sanctions, involvement in internal conflicts and collective security; opposed to the International Criminal Court; aggressive in asserting territorial rights (in Taiwan, Tibet or the South China Sea); and prioritising economic development over political rights, the environment or climate action. When coupled with China's growing economic and military power, and creeping diplomatic and economic influence abroad, China's perceived reluctance to be socialised by international law leads to its ready portrayal as a growing strategic threat, coupled with much angst about the declining fortunes of the United States.

On its part, China resents such criticisms as foreign interference in its sovereign choices, which fails to appreciate China's special historical circumstances, material level of development, and a normatively and procedurally unequal international playing field. As a current Chinese judge of the International Court of Justice writes, "China's adherence to the principle of sovereignty is simply misinterpreted in the west as a disregard of the development of international law, or worse still, considered an excuse to evade its international responsibilities" (Xue, 2007, p. 84).

This article explores China's attitudes towards the regulation of natural resources by international law, whether domestically or at the transboundary and transnational levels. It will consider both the impact of international law on China's own practices and the contribution of China towards shaping international law. At the outset, efforts to understand Chinese attitudes towards the regulation of natural resources are situated within two frames of reference: first, China's experience of international law generally; and secondly, international law's own limited approach to regulating natural resources. The article will suggest that popular or realist conceptions of a relatively isolated, sovereign absolutist China do not accord with contemporary legal realities. While China's construction of strong sovereignty inevitably shapes its attitudes towards legal regulation, practice also suggests that China often adopts a nuanced approach which includes legal compromise, and a commitment to multilateral regulation or bilateral diplomatic settlement of key issues that were hitherto within the competence of national governments. China is often an active and constructive participant in contemporary international law making, even if its socialisation is not necessarily linear (Mushkat, 2011, p. 45), and even if – like all countries, and especially powerful ones – it also seeks to instrumentally use or change international law to secure its own interests and attain a comparative advantage.

This article is foremost grounded in the discipline of international law and its modest objective is to consider the extent to which China's dealings with natural resources can be explained within the accepted modes of international legal reasoning. As such, it does not purport to contribute to international relations theory explaining or critiquing China's behaviour (such as through notions of "compliance" or "cooperation"), which has been considered elsewhere (Chan, 2006; Kent, 2007; Mushkat, 2011). A clearer understanding of China's legal claims, and how they square with international law argumentation, is timely and important against a background of much commentary and controversy about the propriety or otherwise of China's approaches to natural resources.

China and International Law

China's approach to the international regulation of natural resources can only be understood in the context of China's experience of international law generally, which has shaped its contemporary understanding of sovereignty and power relations. China's historical experience of receiving international law is not a positive story. In the nineteenth century, the expansion of European commercial, military and colonial power is perceived to have "unfairly imposed" a "western [international] legal order" on China (Xue, 2005, p. 134), which had hitherto been relatively isolated from deep relations with European powers (Jia, 2010, p. 26). The then law of nations did not forbid the use of military force to pursue foreign policy goals such as colonial expansion, territorial acquisition, commercial exploitation, or even the punishment of a delinquent State. China's encounter with the law of nations brought by European expansion was felt as acute political and cultural humiliation, involving the forcible imposition of unequal treaties; foreign leases, territorial concessions, and commercial and diplomatic privileges; punitive military expeditions; and excessive war reparations (Xue, 2005, p. 134). Japanese imperialism in the 1930s brought further turmoil and little effective protection by international law or institutions. From the nineteenth to the mid-twentieth centuries, then, China experienced international law as a relatively unstructured, laissez-faire system which rewarded the powerful, especially as regards the forcible appropriation of natural resources in foreign territory.

The normative shifts in the postwar United Nations (UN) Charter of 1945 did not immediately alter realities for China, which remained mired in a civil war until the communist victory in 1949. At that time, China terminated existing unequal treaties and sought to re-establish unified sovereign control over its territory. China did not, however, immediately participate fully in the fruits of the postwar order. For a start, mainland China did not assume its seat in the UN until 1971, and it took some time to normalise diplomatic relations with many countries. It was accordingly excluded from key multilateral law-making processes for more than two decades after the war. Further, China's internal difficulties limited its engagement with international law; during the Cultural Revolution (1966–76), "formal legal institutions and legal education were totally abandoned for ten years" (Xue, 2005, p. 135) and relative isolationism marked China's foreign relations. Even though mainland China was not part of the UN until 1971, and had scarcely participated in the creation of international law, China nonetheless indicated its acceptance of much of the postwar international legal order. It did so not only because it had to as a condition of recognition and participation in international social life, but also because "fundamental principles of the legal system as enshrined in the UN Charter reflected certain values they had been fighting for: sovereignty, equality, democracy and self-determination" (Xue, 2007, p. 85). The Charter system was viewed as a shield for weaker States from external interference, and a means of freely pursuing an indigenous system: "a last resort for the developing countries to defend their political system, economic policy, or social stability" (Xue, 2007, p. 85).

Once China was admitted to the UN, its engagement with the development of international law accelerated with participation in major multilateral initiatives such as the 1972 Stockholm Conference on the Environment and the UN Conference on the Law of the Sea. With China's "open policy" of economic reform after 1978, an "ideological emancipation" which challenged traditional views of socialism and development (Xue, 2005, p. 135), China became more and more enmeshed in international legal frameworks and processes and has been part of the evolution of international law to embrace a wider spectrum of interests (Jia, 2010, p. 27). Whereas China's acceptance of international law was initially instrumental – "to increase its international status and promote its interests" – over time its engagement has become "deeper, more meaningful" (Kent,

2006, p. 4). It is currently investing heavily in international law training and education, with the State supporting initiatives such as the Xiamen Academy of International Law and the Xi'an Jiaotong University Silk Road Institute of International Law.

The striking feature of contemporary Chinese foreign policy is not how comparatively aloof China remains from international legal norms, but how quickly it has integrated itself into them. As one Chinese writer notes, "for Europeans, the [Westphalian international legal] system by now is over 360 years old, but for non-European countries, particularly for the Asian and African countries, it is only 60 years old" (Xue, 2007, p. 84).[1] Despite this, China is now a party to almost 300 multilateral treaties (more than 90 per cent of which were adopted in the 30-odd years since 1978) (Xue, 2005, p. 136), and a member of more than 130 international organisations (compared with only 20 in 1978) (Wang and Hu, 2010, p. 194). Further, China has adopted more than 100 bilateral legal cooperation and extradition treaties (Wang and Hu, 2010, p. 195), embraced norms concerning transnational organised crime and anti-corruption (Wang and Hu, 2010, p. 199), and acceded to membership of the World Trade Organisation. China's approach is encapsulated by its doctrine of the "harmonious world": observing UN Charter principles and purposes, complying with international law, and promoting democracy, harmony and "win-win" cooperation in international relations (Wang and Hu, 2010, p. 197).

China's embrace of international law cannot of course be "proxied" by its formal treaty commitments (Mushkat, 2010, 518). China's domestic law too has been increasingly influenced by international standards in areas such as public and administrative law, criminal law, judicial cooperation in civil and commercial matters (Xue, 2005, p. 136–37; Jia, 2010, p. 42), and (as discussed below) world trade law. International treaties occasionally take priority in some areas of law, and unclear domestic laws may be interpreted consistently with international obligations (Gao, 2007, pp. 18–19; Wang and Hu, 2010, p. 194; Jia, 2010, p. 44). "Soft" international norms have also influenced China's domestic policy, including on sustainable development, poverty alleviation, gender, food security and the Millennium Development Goals (Xue, 2005, p. 137).

More controversially, China has maintained distinctive legal positions at variance with those of certain liberal States. It has shown little enthusiasm for UN support for democratisation (not, incidentally, an obligation of States under international law). Further, while committing to universal human rights – and China has signed most treaties (Xue, 2005, p. 136) – China has warned that civil and political rights should not come at the expense of economic, social and cultural rights (Chinese Government, 2005, p. 697), or interfere in domestic affairs. China has often resisted international supervision of human rights (for instance, by the International Labour Organisation or UN treaty bodies), arguing that it is an infringement of its sovereignty (Kent, 2006, p. 10). Many other States (including western ones) have also had tense relationships with global human rights bodies for similar reasons, and China's attitude is a matter of degree, not of kind. Chinese human rights protection will remain limited until there is genuine political reform (Wan, 2007, p. 728). Nonetheless, the domestication of rights has changed behaviour by introducing formal legal and administrative constraints, structuring political discretion, raising public awareness, and raising the costs of non-compliance. China has also participated in bilateral dialogues on human rights (for instance, with Australia) (Van Ness, 1992; Kent, 2001; Fleay, 2008), which have provided an alternative means of socialisation, albeit one more malleable and limited than supervision by multilateral mechanisms.

China's growing power has brought with it China's own awareness of its role in contributing to international public life. China is active in UN peacekeeping (Xue, 2007, p. 88), with 12 missions in Africa since 1990 involving more than 4,000 people. It has actively contributed to debates about UN reform, arguing for stronger representation of developing countries in the Security Council; strengthening the General Assembly as a democratic body; and reinvigorating the UN Economic and Social Council's role in development (Chinese Government, 2005, p. 697). On collective security, China has cautiously endorsed attention to "new" security threats or "human security" issues (Chinese Government, 2005, p. 697; UN General Assembly, 22 April 2005), but called for "reversing the trend" of the UN prioritising security over development in a post-9/11 world, noting that development is the "bedrock" of collective security (Chinese Government, 2005, p. 697). China has also cautioned against "reckless" interventions, and argued for prudent, collective control of security measures (Chinese Government, 2005, p. 694) – understandable after the illegal aggression against Iraq of 2003. It is not a party to the International Criminal Court (ICC), and has warned against an overly powerful ICC, which would be "likely to be abused and politicised by powerful States … and would thus become an instrument for violating the rights and interests of small countries under the pretext of human rights and justice" (Wang and Hu, 2010, p. 200). At the same time, China's practice is not dogmatic; for instance, it did not veto the Security Council's referral of Sudan and Libya to the ICC, or veto international military intervention in Libya, or UN and African Union deployments in Sudan. Its formal adherence to strong sovereignty has given way to pragmatic responses to individual humanitarian crises, where the host State consents or there is Security Council authorisation (Davis, 2011). Its positions can, however, be inconsistent; in 2011 China invited President Bashir of Sudan to visit China, despite acquiescing in the Security Council's earlier referral to the ICC. China has also taken a very cautious approach to universal criminal jurisdiction (Chinese Government, 2011b). It resisted international efforts to deal more firmly with the Assad regime during the Syrian civil war of 2011–12, or to pressure Sri Lanka on accountability for civil war crimes.

Despite the common depiction of a resource-hungry China coddling nasty regimes such as Sudan, the reality is more nuanced: a leading empirical study shows that there is little correlation between China's resource interests and its position on security matters in the Security Council (Houser and Levy, 2008). Indeed, China has been far less obstructive generally of the UN Security Council than other permanent members: between 1971 and 2006, the United States vetoed 76 resolutions to China's four (Houser and Levy, 2008, p. 66). Consensus between China and the United States has grown on the Council in recent decades (Houser and Levy, 2008, p. 64). It has also been actively involved in multilateral arms control efforts concerning nuclear, chemical and biological weapons, and North Korea (Kent, 2006, p. 5), but resistant to controls on land mines, cluster munitions and conventional weapons.

This brief survey of China's engagement with international law reveals that China's historical suspicion of international law (as a tool of colonialism) gradually gave way to endorsement of key features of the modern law (including sovereign equality, non-interference, multilateralism, peaceful resolution of disputes, democratisation of international governance, and the international rule of law) (Jia, 2010, p. 34). It also comes as no surprise that "certain publicly held positions of China are shared by States, in general" (Jia, 2010, p. 60), including, for instance, a strong notion of sovereignty, and a preference for diplomatic settlement of disputes over binding adjudication. China has not accepted the compulsory jurisdiction of the International Court of Justice (ICJ) under article 36(2) of the ICJ Statute. As is well accepted, however, submission to ICJ jurisdiction is hardly the litmus test of whether a State takes international law seriously: Austinian command and control theories of "law" went out of fashion half a century ago. Around two-thirds of all States (including the US, Russia and France) have not submitted to the ICJ's jurisdiction; and judicial settlement of transnational disputes is not always appropriate. At the same time, China is hardly leading from the front in supporting the UN's primary judicial organ – one may expect more leadership from powerful States – though Chinese judges are routinely on the Court and one (Judge Shi) has served as President.

Overall, while China was largely a "responsive rule-taker during the formation of the existing international legal order" (Qi, 2008, p. 337), it has also attempted to improve unjust elements of it (Wang and Hu, 2010, p. 200), and is increasingly active in law-making. The next section of this article considers China's engagement with international law in relation to the regulation of key areas of natural resources, in light of China's historical experience and its conception of sovereignty.

China, International Law and Natural Resources

Territorial resources: Land and sea

China has been much criticised for its perceived aggressive territorial acquisitions or ambitions in Tibet, Taiwan and the South China Sea. All States have a vital sovereign interest in securing their territorial integrity and China is no exception. Effective control over national territory is, after all, an incident of statehood under international law (Montevideo Convention, 1933, Article 1). China's approach to safeguarding its territorial resources can largely be explained within a conventional international law framework. There is little evidence that China has pursued expansionist territorial ambitions beyond the level to which it believes it is entitled under international law, even if its interpretation of the law is contested in some cases. China also tends to pursue peaceful rather than forcible settlement of its disputes.

For China, the flashpoints of Tibet, Taiwan and the South China Sea can all be explained by a proper analysis of China's title to territory under international law. Tibet has long been considered historically part of China, and there is considerable historical evidence to support that position. While Tibet exercised considerable autonomy at various times, it was not treated as an independent State prior to the Chinese "invasion" in 1951 (Crawford, 2007, p. 325) – even if its status was largely determined by external powers. Militant resistance to China in the 1950s, and later political efforts, failed to establish Tibetan secession. While there are plausible arguments in favour of Tibet's historical independence, the international community uniformly endorses China's position. There is negligible support from other States or the UN for Tibetan independence and statehood, and the international community does not regard Tibet as either foreign occupied territory or involving a "people" entitled to exercise self-determination – as evidenced, for instance, by an absence of United Nations General Assembly or Security Council resolutions to that effect. Tibet is treated by the international community as the sovereign territory of China. Even the Dalai Lama and the Tibetan Prime Minister do not claim independence for Tibet, as opposed to genuine internal autonomy. China is, however, evidently responsible for serious and systemic violations of human rights in Tibet, including the repression of the political and religious freedoms of minorities and arbitrary detention. Some of its legal positions are also unsupportable expressions of paranoia, such as its declaration that foreign politicians are interfering in China's internal affairs and infringing its sovereignty by meeting with the Dalai Lama (Chinese Foreign Ministry, 2011a).

China's position on Taiwan is even stronger under international law. At first sight, Taiwan appears to operate as a State, with an effective government exercising control over a permanent population and territory, and entering into agreements with other States. Yet territorial Taiwan is not recognised by the UN or any country as an independent sovereign State (Crawford, 2007, pp. 198–221). Taiwan itself has not consistently claimed that territorial Taiwan is an independent State separate from mainland China, but has historically (and ineffectively) claimed to be the government of a single unified Chinese State. Most States deal in formal terms with (the People's Republic of) China as the relevant State, even if they maintain practical diplomatic relations with the Republic of China (Taiwan) (and only a handful of small States (about 23) recognise Taiwan as the government of the Chinese State as a whole). Most States and international organisations deal with Taiwan as a *sui generis*, limited legal entity for specific, practical purposes, for instance as a fishing or aviation entity (Crawford, 2007, p. 220). China's "One China" policy and its view of the status of Taiwan (as part of Chinese territory) are generally accepted, though there is disagreement over the terms (such as the extent of autonomy) and means (by war or peace) of reunification.

In strict terms, Taiwan is the defeated rump of an earlier Chinese nationalist government deposed in a civil war. Under international law, the newly established communist government of mainland China, initially de facto (from 1949) and de jure (from 1971) may even be entitled to forcibly suppress an insurgency in its own territory. China's designs on Taiwan are not about territorial expansion or resource acquisition, but a conventional legal attempt to unify governance of sovereign territory. China might be understandably puzzled at the US's defensive strategic posture over Taiwan, given the US acquiescence in China's more contestable claim to Tibet.

On both Tibet and Taiwan, it may be that China presents interpretations of factual and historical situations which best suit its case. Others may contest those claims, and the evidence is both mixed and unsettled by adjudication. But China's views by and large are not only plausible, but reasonable (in international law terms, which does not always accord with substantive justice), and supported by many States and international jurists. Far from reflecting a "land grab" or a naked appropriation of foreign resources, China has been careful to articulate its claims within an international law framework. It has also contributed reasoned and principled legal views to external judicial processes which touch upon the substantive issues at stake in both Tibet and Taiwan. Recently, before the International Court of Justice in the *Kosovo Advisory Opinion* case, China argued that Kosovo's declaration of independence was contrary to the principle of State sovereignty and territorial integrity, and that remedial self-determination was not available (quoted in Jia, 2010, p. 46). It also argued that Kosovo's claim was inconsistent with the negotiated mode of settlement stipulated by the relevant Security Council resolution. China's legal position was shared by quite a few States, and reflects a conventional and widespread analysis. The Court itself did not license remedial self-determination, finding only that no international rule prohibits unilateral declarations of independence. The latter view is itself controversial, and not in accordance with the views of a significant number of States.

In other situations, China has often utilised peaceful methods of resolving territorial disputes. It has, for instance, completed peaceful border demarcation negotiations with Russia, Central Asian countries and Vietnam (Wang and Hu, 2010, p. 195). It is true that China fought a war with India in 1962 over disputed Himalayan borders. But that conflict (and before it, China's intervention in the Korean War in 1950 – fighting *against* UN forces) cannot be seen as an aggressive or acquisitive war of territorial expansion, as it was seen at the time by western eyes preoccupied with containing communism. Rather, it involved a more conservative Chinese effort to assert and defend sovereignty over presumed Chinese territory, in a factually complex situation marked by unclear historical titles and confusion on the ground about India's intentions. It is striking that the war did not escalate beyond the frontier, demonstrating China's limited goals; and today China is peacefully negotiating with India to resolve outstanding border issues (Chinese Foreign Ministry, 2011b).

China's conventional understanding of sovereignty as unified territorial control has, however, been tempered by a certain pragmatism in practice. With the end of British and Portuguese colonialism in Hong Kong and Macao respectively, China's innovative "one country, two systems" doctrine preserved distinctive, relatively autonomous legal regimes in those territories, including towards their prior treaty obligations (Xue, 2005, p. 139). Such creative flexibility was not required by international law, since sovereignty over national territory does not come encumbered with obligations to confer internal autonomy in postcolonial situations (in contrast, for instance, to minority or indigenous regimes). China's progressive approach has struck a "balance between important values of sovereignty and the ideal of the rule of law" (Jia, 2010, p. 61) by preserving "extra-sovereign" values there which are not part of life in the same way, or to the same extent, on the mainland. Maritime disputes in the South China Sea remain prominent and unresolved. The Chinese view is that "China has simply been safeguarding what's rightfully her own", based on good historical title over other claimants (Shen, 2002, p. 157). China's claims there are undoubtedly driven by resource interests and strategic defence, but they remain underpinned by arguments made by reference to principles of international law, even if some claims are more tenuous than others. The international law of the sea, and of title to territory, do not give easy answers to many of the maritime disputes involving China (Hsiung, 2007, pp. 136–48; Ramos-Mrosovsky, 2008; Manjiao, 2011), both because of uncertainties in the law and hard disputes about the facts. Maritime disputes involving competing, distant and often tenuous historical claims are notoriously difficult to determine; the flaring dispute with Japan over the Senkaku/Diaoyu Islands is a case in point, with credible claims on both sides (Schoenbaum, 2008, pp. 45–47; Shaw, 2008). Further, the periodic, isolated military confrontations over island territories are not solely attributable to China, but also flow from the robust assertion of competing claims by various other States. The maritime disputes are, however, dangerously fuelled by popular nationalism in China (and other States), and embedded in wider political, cultural and historical enmities with its neighbours (Suganuma, 2007), which makes Chinese behaviour unpredictable as it reacts to domestic constituencies.

Far from seeking to resolve disputes mainly by force, however, China has committed to ongoing maritime boundary discussions (Wang and Hu, 2010, p. 195), including through the 2002 *ASEAN Joint Declaration and Code of Conduct*, aimed at maintaining the peace and peacefully settling disputes through negotiations, and the 2011 *Agreement on Basic Principles Guiding Settlement of Sea Issues* with Vietnam. China has also negotiated fisheries agreements with Japan and South Korea, and seabed joint development with Japan (pending delimitation) (Gao, 2008), signifying a preference for practical cooperation over legal confrontation. As the most powerful State in the region, negotiated outcomes will necessarily benefit China (Duong, 2006) more than independent, adjudicated settlements, but its approach is within the accepted modes of consensual dispute settlement under international law. Negotiation may even be preferable in some situations because it would produce cooperation and workable compromise (Wu and Zhang, 2010, pp. 148–49) rather than winners and aggrieved "losers" in adjudication (Schoenbaum, 2008, p. 25). By contrast, China has more forcefully defended its controversial legal demand for prior authorisation of foreign warships transiting through its territorial sea (Zou, 2005), resulting in provocative incidents involving US warships. The US objects that such a requirement infringes upon the protected freedom of navigation under international law (Franckx, 2011, p. 191). The Chinese position is *prima facie* inconsistent with the UN Convention on the Law of the Sea 1982 (UNCLOS) to which China is a party, although China has reserved its legal obligations under that treaty.

It is nonetheless significant that China's assertion of a strong sovereignty over its maritime territory is cast in elaborate legal terms, and does not take the form of simple rule-breaking. Moreover, China is a party to UNCLOS, whereas the principal objector to its legal stance, the US, is not, though (like many States) China is reluctant to utilise its binding dispute resolution mechanisms. It is also hard to imagine the US viewing favourably any future Chinese practice of transiting its blue navy warships through US waters – as the US does in Chinese waters – and which may ultimately provoke a change in China's own position over time, as it gains capabilities to project its naval power.

A review of China's attitudes towards territorial resources suggests that, for the most part, China accepts and espouses a fairly conventional understanding of international law, which largely accords with its preference for strong sovereignty. There is little evidence of China as a territorially expansionist, resource acquisitive law-breaker. China's behaviour concerning Tibet and Taiwan can be analysed consistently with a fairly mainstream international law approach, whereas in the South China Sea it is perhaps pushing the boundaries further. Peaceful bilateral negotiation and multilateral governance have featured more prominently than forcible solutions. Normatively, China's strong sovereignty has occasionally ceded ground, as in the decentralised approach to Hong Kong and Macao. Farther afield, it is also noteworthy that China accepts certain international law doctrines which are predicated precisely on a rejection of strong national sovereignty over resources. For instance, under the Law of the Sea,

China accepts the notion that the deep seabed is the common heritage of mankind [humanity] (Chinese Government, 2011a); and endorses the special international regime governing Antarctica, which suspends sovereign territorial claims and precludes resource extraction (Chinese Ambassador to Argentina, 2011).

The historical evidence suggests, therefore, that China's emphasis on strong sovereignty is a defensive rather than aggressive concept, one which is geared towards reclaiming or defending sovereign territory and resources, rather than directed outwards to acquiring those of others. China's growing power, and the increasing needs of its population, do not necessarily change that historical equilibrium by bringing demands for territorial expansion or forcible resource acquisition. In a modern economy where secondary and tertiary industries are important, and primary industries such as land are less significant, "intensive development through economic growth [and trade] is generally preferable to military and extensive expansion" (Rosecrance, 2006, p. 33). China is heavily integrated into and dependent on the global economy and it is in China's interests not to disrupt those beneficial arrangements (Rosecrance, 2006, pp. 34–35). As discussed below, China's foreign investment strategies in Africa and elsewhere enable it to acquire the resources it needs on market terms. Moreover, the military, economic and political costs of acquiring and holding foreign territory are exorbitant (Rosecrance, 2006, pp. 32–33) and a strong disincentive. Securing its needs through multilateral economic cooperation is a far more efficient and desirable means than territorial expansion, even in a world of resource scarcity. In addition, as a member of the Security Council, China is socially constrained by relatively robust multilateral institutions which curb tendencies towards aggression.

Permanent sovereignty over natural resources

As Anghie observes, "[t]he end of formal colonialism ... did not result in the end of colonial relations" (2008, pp. 44–45), because of the persistence of international economic laws which preserve the economic dependency of newly decolonised States, whether through unequal trading arrangements or foreign commercial control over natural resources (for instance, through long-term resource concessions on unfavourable terms). One of the achievements of the postwar international legal order was recognition of the right of independent States to assert permanent sovereignty over natural resources, in the interests of national economic development (Jia, 2010, p. 18).

Conceptually central to the economic sovereignty of developing countries were the 1962 General Assembly Resolution on Permanent Sovereignty over Natural Resources (UN General Assembly Resolution 1803 (XVII) (1962)), and the subsequent 1974 Charter of Economic Rights and Duties of States (UN General Assembly Resolution 3281 (XXIX) (1974)), to which China has been consistently committed (Jia, 2010, pp. 19–21). Such norms were consistent with both socialist and developmental agendas in China. Further critical efforts by developing countries to construct a New International Economic Order (to balance the market's emphasis on efficiency with other values such as fairness and distributive equity) (Gathii, 2008, p. 255) and a "right to development" (the latter still supported by China) were not successful. Even so, various substantive and procedural concessions were extracted along the way which moderated some global inequalities. Utilising its natural and human resources, China has lifted hundreds of millions of people out of poverty and sustained an average annual growth rate of more than 9 per cent over the 25 years since 1980 (Xue, 2005, p. 137). China's utilisation of resources has, of course, been socially uneven in its dividends, involved violations of civil and political rights, and led to environmental degradation.

Domestic, transboundary, and transnational regulation of resources

The growth of international environmental law partly moderates the pro-development principle of permanent sovereignty over natural resources. Through the principle of sustainable development, the domestic use and management of natural resources is elevated beyond exclusive sovereign, domestic jurisdiction to an international concern and subject to international regulation, even if uncertainty remains about its scope and its relationship to human rights and international economic law (Birnie and Boyle, 2002, p. 85). The subsidiary principle of sustainable utilisation, use and conservation of natural resources, and the precautionary principle, aim to preserve scarce resources and challenge the classical approach to resources as infinitely exploitable.

China has formally adopted the principle of sustainable development, as is evidenced by official statements such as this:

> China is a responsible country. Our government sticks to the path of sustainable development, steadfastly pursues the policy of reducing resource consumption and preventing environmental pollution, and works energetically to build a society that is resource-efficient and environment-friendly (Ma, 2008, p. 569).

Since the 1980s, China has adopted strong domestic environmental laws, but enforcement in practice remains poor for many reasons (Beyer, 2006, p. 185). There is also increasing interest in "corporate social responsibility" in China (Lin, 2010), including laws and guidelines.[2] The problem is not a normative rejection of environmental norms, but their domestic implementation. China's attitude towards transboundary environmental governance is more mixed. China actively complied with global efforts concerning ozone protection (Kent, 2006, p. 8). Certain transboundary harm emergencies, such as the Song Hua River pollution incident, have been dealt with according to the established principles of international law (Wang and Hu, 2010, p. 195). There is some evidence of China's cooperative participation in transboundary river management, as in "soft" cooperative regulatory approaches to the Tumen River (involving China, Russia, Mongolia and the Koreas) (Marsden, 2010). Less successful is China's approach to the Mekong River Basin, which originates in Tibet. China is not a member of the Mekong River Commission, and China's upstream development of hydropower dams has been pursued with relative disregard for less powerful downstream riparian States in Southeast Asia. Chinese officials have protested against external interference in China's exploitation and utilisation of its water resources (Ma, 2007, p. 769). China's attitude is not atypical, however, given that control over scarce water resources is a sovereign concern globally and few States have leapt at the opportunity to strengthen multilateral governance by adopting the *UN Transboundary Watercourses Convention*. At the same time, China has demonstrated progressive instincts in some law-making processes. For example, in the International Law Commission's work on State responsibility for transboundary harm from hazardous activities, China sought to widen State responsibility (Ma, 2008, p. 568). In other environmental areas, China has played a creative role in brokering new norms. China's support for the principle of "common but differentiated responsibilities" (Wang and Hu, 2010, p. 198) was important in securing global agreement on climate change. The principle recognises the historical responsibility of developed countries for carbon pollution, while enabling developing countries to "catch up" without bearing an unfair burden for mitigating emissions too soon. China has made considerable efforts to reduce its carbon emissions and energy intensity (Meidan, 2010, pp. 308–13), without agreeing to binding reductions targets (Kent, 2006, p. 9), even if its efforts are as much directed to reducing air pollution, improving energy security, and enhancing economic productivity. Its commitments at the policy level are not always matched by implementation, however, and measures announced thus far will not stop the growth in emissions (Garnaut et al., 2008, p. 182).

While the principle of differentiated responsibilities arguably aligned with China's reality at Kyoto in 1997, the reality by Copenhagen in 2007 was quite different. Not only is China's economic position now much stronger, and its carbon pollution much higher (now the world's largest emitter), but China's approach at Copenhagen divided developing countries, many of whom no longer regard themselves in the same class as a richer, higher polluting China. It remains to be seen whether China adapts its negotiating position to its new status – or whether it continues to instrumentally rely upon a principle now ill-suited to its changing responsibilities. Recently China protested, for instance, the legality of European Union measures to address climate change which impact on the Chinese aviation industry (Chinese Civil Aviation Administration, 2011). In addition to domestic and transboundary environmental regulation, one further area deserves mention. Chinese State companies and private investors are increasingly doing business in foreign States, particularly in regards to China's rapidly growing engagement with natural resources in Africa (Zweig and Bi, 2005) and Myanmar (especially in gas, oil, hydrocarbon and related transport and trade infrastructure). Foreign corporate activity is not directly regulated by international law. There is no international treaty requiring States to regulate the extraterritorial activities of their companies; States have been reluctant to view their human rights obligations as requiring the regulation of companies abroad; corporate social responsibility is often voluntary and non-binding; and there are few forums in which those affected by corporate or investor harms can seek remedies.

Local legal systems in some of the destinations for Chinese investment are often inadequate because of poor regulatory controls, weak enforcement, corruption or vulnerability to Chinese business or diplomatic pressure. At the same time, some countries increasingly prefer to deal with Chinese companies precisely because they come with fewer strings attached. The insistence of the World Bank, Asian Development Bank, and western governments on compliance with social, environmental or rights-based conditions has in some cases resulted in their displacement by Chinese companies and a loss of influence by those actors, even if the effectiveness of conditionality may be overstated (Alden and Alves, 2009, p. 19). Chinese exploitation of resources is thus causing potentially greatest harm not in China or its neighbours, but rather through transnational operations further afield. Even so, there are signs of change, as China responds to international and local pressures to improve its practices. From 2007 onwards, the State Council and various ministries have increasingly sought to regulate Chinese foreign investment; the Export-Import Bank of China has issued environmental and social impact policies; and there are agreements between China and the International Finance Corporation on environmental matters, and the World Bank on exchanging project information (Alden and Alves, 2009, p. 20). Despite progress in standard setting, there remains a lack of enforcement on the ground (Alden and Alves, 2009, p. 20). The adoption of the *UN Guiding Principles on Business and Human Rights* in 2011 may gradually help to change the behaviour of Chinese entities doing business abroad. The UN Human Rights Council resolution 17/4 (2011) endorsing the *Principles* was adopted without a vote, implying China's acceptance of them as a Council member. The former UN Special Representative for Business and Human Rights, John Ruggie, notes that China understands that poor corporate behaviour abroad reflects badly on it, and its companies are on a steep learning curve in improving operations (Business and Human Rights, 2011, p. 119). There has also been push back from affected States, with Myanmar suspending a Chinese dam project on the Irrawaddy River in 2011, even if Chinese exploitation of Burmese resources is a substantial problem, particularly in the border regions (Shee, 2002, p. 44).

China's engagement in Africa illustrates some of these tensions and trends. On the one hand, Chinese investment in African natural resources, particularly energy and mining, has been criticised for not respecting environmental and labour standards; using Chinese instead of local workers; and corruption (Alessi and Hanson, 2012). China's preferred method of trading infrastructure for resources, typically on a bilateral basis (Alden and Alves, 2009, p. 9, p. 10, p. 16), does not always optimise the developmental dividend for an African State. China has been more interested in profit and less interested in governance reform, and its interests in Africa's primary commodities replicates earlier (and continuing) neo-colonial relations with western States and companies (Alden and Alves, 2009, p. 18). At the same time, Chinese capital has "emboldened" some African States to pursue policies which would be opposed by international institutions or the West (Alden and Alves, 2009, p. 18). The picture is, however, more nuanced than is often portrayed. African States have often been willing participants in this mode of development, not least because it diversifies their investors and counterbalances the eco-political dominance of the US and EU (Alden and Alves, 2009, p. 9). China's involvement in Africa is not new, but a deepening of old geopolitical ties cemented since the 1960s, when China supported African liberation movements and developmental aspirations (Alden and Alves, 2009, pp. 7–8). Recently, some States, such as Angola, have grown by utilising Chinese investment while simultaneously pursuing governance reform and improving transparency (Alden and Alves, 2009, pp. 18–19). China's provision of "hard" infrastructure has been beneficial in many States, as has the expansion of Chinese aid and development assistance. China can also be responsive to international pressure to be a "responsible stakeholder", including through international efforts to enhance governance, transparency and sustainability of resource development (Alden and Alves, 2009, p. 20, p. 21), even if there is a long way to go.

Economic integration

Another key frame of reference in China's dealings with resources concerns its participation in world trade, and its approach to resolving commercial disputes. Some Chinese critics observe that China's entry into the World Trade Organisation (WTO) in 2001 was under "exceptionally unfavourable, non-reciprocal and asymmetric terms of membership" (Wu, 2011, p. 227), including less than full treatment as a developing country. Certainly its accession was a site of political leverage for powerful actors such as the US and EU. Even so, China's WTO accession has brought considerable economic benefits to China and helped it to become the world's largest exporter, including through "most-favoured-nation" status with the US (Wu, 2011, p. 237).

WTO membership is attractive because it brings "stability, certainty and predictability" in China's trade relations. At the same time, "the WTO framework respects the State sovereignty of its members, and leaves it to internal domestic policy to control national security, environmental protection, redistribution of wealth, public morals and culture" (Wu, 2011, p. 232). The WTO was not designed to realise the theories of those who believe that economic integration inevitably brings political democratisation, and such transformative consequences are modest in China to date (Guo, 2008, pp. 354–55). WTO membership nonetheless involves considerable sovereign concessions in exchange for trade benefits. First, there is a dramatic internal reform aspect to WTO membership: since 2001 China has modified 3,000 domestic laws and regulations (Xue, 2005, p. 138) and up to two million local regulations (Guo, 2008, p. 344) to comply with its WTO obligations, suggesting a deep reshaping of domestic laws. Implementation is more problematic, particularly at the local level and in the absence of an independent judiciary (Guo, 2008, p. 343). But an important feature of internal reform may be a gradual rule of law dividend, with increased regulatory transparency, public consultation in law-making, and administrative review of governmental action, sometimes with spill-over effects into law and legal institutions more generally (Wu, 2011, p. 233).

Secondly, as a condition of membership, China submitted to the binding dispute resolution mechanisms of the WTO, including in cases brought by the US, largely due to the reciprocal economic interests involved in being a good economic citizen. China is participating actively in the procedures when claims are brought against it, and after a slow start, is bringing its own claims. By June 2012, China was complainant in eight cases, respondent in 26 cases, and a third party in 89 cases (WTO, 2012). China has enjoyed success in claims against the US (for example, in *United States—Definitive Antidumping and Countervailing Duties on Certain Products from China*, 2011) giving it confidence in the WTO procedures. At the same time, claims brought against it in controversial areas provide a structured means of defusing escalating trade disputes, as in the case brought by the US, EU and Japan in March 2012 against China's export quotas on rare earth minerals (*China – Measures Related to the Exportation of Rare Earths, Tungsten and Molybdenum*, 2012), and which China partly defends on environmental grounds (Gu, 2011, p. 774). That case followed China's loss in a raw materials case brought by the EU (*China – Measures Relating to the Exportation of Various Raw Materials*, 2011), which led to China reforming its practices in order to comply with the ruling. China's submission to binding dispute settlement is significant, given that in commercial disputes before national courts, it is increasingly isolated as one of the few States to insist on a doctrine of absolute immunity for State companies before foreign courts (Qi, 2008, p. 326). China has also recently resorted to the International Centre for Investment Disputes for the settlement of that category of commercial disputes (Heymann, 2008).

Like many countries, China has increasingly pursued bilateral free trade agreements (FTAs) alongside the multilateral WTO system. By May 2012, China had concluded nine FTAs, with a further five under negotiation and four feasibility studies under way; 31 countries are included in these arrangements (Chinese Ministry of Commerce, 2012). On the one hand, this might be seen to weaken the multilateral trading system, by going outside it and creating a differentiated web of bilateral arrangements. But FTAs are a lawful and common device for expanding a State's trade opportunities, and in China's case were initially borne out of frustration at the slow pace of WTO accession. They are also an understandable response to the unequal terms of China's WTO membership, and allow China to enjoy greater bargaining power and diversify its trading relationships (Gao, 2007, pp. 27–28). Its FTA with ASEAN has significantly boosted trade and economic activity in Southeast Asia (Xue, 2012, p. 208).

WTO membership is also an opportunity for China to express its sovereign preferences by working to restructure the world trade system. Within the WTO China has played a constructive role in seeking to improve the fairness of trade, specifically by the elimination of agricultural subsidies and reducing protectionism (Chinese Government, 2005, p. 687). Liberalisation of trade in agriculture has floundered since the Doha Round, even though it is widely agreed that it is an essential component (and a plank of the Millennium Development Goals) of a market-based approach to economic growth and development. China has also argued for reform of the international financial system to make it more equal and mutually beneficial (Chinese Government, 2005, p. 687); third world debt relief; and the strengthening of South–South cooperation as an alternative framework of cooperation. China's historical experience as a recipient of foreign aid, and now increasingly as a donor, gives it particular authority to also speak on development assistance, as a policy-oriented "soft" process of resource redistribution. Thus China has described freedom from want as the "most urgent task" for the UN (Zhang, 2005) and emphasised the importance of achieving the Millennium Development Goals (Wang, 2005). Balancing its approach to strong sovereignty and its desire to contribute as a good international citizen, China supports the UN national target for Official Development Assistance of 0.7 per cent of GDP. China argues, however, that development assistance should be more sensitive to national conditions, including by respecting the participation and autonomy of recipient countries (Chinese Government, 2005, p. 687).

Conclusion

International law has slowly grown on China, from difficult beginnings (in which law was an apologia for foreign exploitation) through a rapid socialisation into largely beneficial norms from the 1980s onwards. China is now far more law-abiding than law-breaking, including as regards its defensive territorial claims, and is increasingly enmeshed in thicker global social relations. It accepts the basic elements of the international legal order and UN system, and has internalised key rules which suit its interests (such as permanent sovereignty over natural resources, sustainable development, or world trade obligations).

It is no surprise that China embraces international law when some of it is conservatively State-centric rather than transformative of social and power relations. China has interpreted some norms (such as sovereignty) or facts (such as historical ties to territory) according to its own understandings; worked to modify or block norms which do not suit it or with which it is out of step (as in the navigational rights of warships, restrictive immunity, or aspects of UN reform); and cooperatively developed norms to address common new situations (as with aspects of transboundary harm). Along with many States, it has sometimes been rather inert on issues where there are convenient regulatory lacunae (such as in corporate activity abroad) or weak institutions (as in transboundary river governance). Practice reveals that China is also less dogmatic than is commonly thought and has embraced creative compromises (as in Hong Kong and Macao, or in climate change policy) to balance competing legal interests.

On the whole, China takes international law seriously (Wang and Hu, 2010, p. 200), and views sovereign equality and multilateralism – including the protection of the weak from the strong – as positive social values. That is a good sign in a world where China is rising, and resources are becoming scarce. It is also cause for optimism, in contrast to the bleak narrative of future superpower competition between China and the US, unrestrained by international law, which some predict (Posner and Yoo, 2006).

Notes

1. That observation underestimates the extent to which international law did, in fact, deeply shape relations (positively and negatively) between European and non-European powers prior to 1945, and between non-European powers, but it illustrates a dominant perception amongst Chinese elites.
2. Including Article 5 of the 2006 Chinese Company Law (requiring companies to "undertake social responsibility"); the Chinese Ministry of Commerce's 2008 *Guidelines on Corporate Social Responsibility for Foreign Invested Enterprises*; and the Chinese Academy of Social Sciences' (Research Centre for Corporate Social Responsibility) 2009 *Blue Book Report on Corporate Social Responsibility.*

References

Alden, Chris and Anna Christina Alves (2009) China and Africa's natural resources: The challenges and implications for development and governance. Occasional Paper No. 41, South African Institute of International Affairs, September.

Alessi, Christopher and Stephanie Hanson (2012) Expanding China-Africa oil ties. *Council on Foreign Relations Backgrounder*, 8 February.

Anghie, Antony (2008) The evolution of international law: Colonial and postcolonial realities, in Richard Falk, Jacqueline Stevens and Balakrishnan Rajagopal (eds), *International law and the third world: Reshaping justice*, pp. 35–50 (Oxon: Routledge-Cavendish).

Beyer, Stefanie (2006) Environmental law and policy in the People's Republic of China. *Chinese Journal of International Law* 5(1), pp. 185–211.

Birnie, Patricia W. and Alan E. Boyle (2002) *International law and the environment* (Oxford: Oxford University Press).

Chan, Gerald (2006) *China's compliance in global affairs: Trade, arms control, environmental protection, human rights* (New Jersey: World Scientific).

China – Measures related to the exportation of rare earths, tungsten and molybdenum (Complainants: Japan, EU, US) (2012) DS433, 432, 431, filed 13 March.

China – Measures relating to the exportation of various raw materials (2011) WT/DS394/R, WT/DS395/R, WT/DS398/R, 5 July.

Chinese Ambassador to Argentina (2011) Statement on the Antarctic, 23 June, in Lijiang Zhu (ed.), (2012) Chinese practice in public international law: 2011. *Chinese Journal of International Law* 11, paras. 1–77, para. 17.

Chinese Civil Aviation Administration (2011) Joint statement on EU's inclusion of aviation into EU emissions trading scheme, 27 September, in Lijiang Zhu (ed.), (2012) Chinese practice in public international law: 2011. *Chinese Journal of International Law* 11, paras. 1–77, para. 64.

Chinese Foreign Ministry (2011a) Statement on President Obama's meeting with the Dalai Lama, 16 July, in Lijiang Zhu (ed.), (2012) Chinese practice in public international law: 2011. *Chinese Journal of International Law* 11, paras. 1–77, para. 13.

Chinese Foreign Ministry (2011b) Statement on China-India border issue, 15 September 2011, in Lijiang Zhu (ed.), (2012) Chinese practice in public international law: 2011. *Chinese Journal of International Law* 11, paras. 1–77, para. 14.

Chinese Government (2005) Position paper of the People's Republic of China on the United Nations reforms. *Chinese Journal of International Law* 4(2), pp. 685–98.

Chinese Government (2011a) Statement to the UN General Assembly, 5 October, in Lijiang Zhu (ed.), (2012) Chinese practice in public international law: 2011. *Chinese Journal of International Law* 11, paras. 1–77, para. 19.

Chinese Government (2011b) Statement to the UN General Assembly, 12 October, in Lijiang Zhu (ed.), (2012) Chinese practice in public international law: 2011. *Chinese Journal of International Law* 11, paras. 1–77, para. 48.

Chinese Ministry of Commerce (2012) China FTA network. http://fta.mofcom.gov.cn/english/index.shtml, accessed 29 May 2012.

Crawford, James (2007) *The creation of states in international law*, 2nd edition (Oxford: Clarendon).

Davis, Jonathan (2011) From ideology to pragmatism: China's position on humanitarian intervention in the post–Cold War era. *Vanderbilt Journal of Transnational Law* 44, pp. 217–83.

Duong, Wendy (2007) Following the path of oil: The Law of the Sea of realpolitik – What good does law do in the South China Sea territorial conflicts? *Fordham International Law Journal* 30, pp. 1098–208.

Fleay, Caroline (2008) Engaging in human rights diplomacy: The Australia-China bilateral dialogue approach. *The International Journal of Human Rights* 12, pp. 233–52.

Franckx, Erik (2011) American and Chinese views on navigational rights of warships. *Chinese Journal of International Law* 10(1), pp. 187–206.

Gao, Jianjun (2008) Joint development in the East China Sea: Not an easier challenge than delimitation. *Marine and Coastal Law* 23, pp. 39–75.

Gao, Henry (2007) China's participation in the WTO: A lawyer's perspective. *Singapore Year Book of International Law* 11, pp. 1–34.

Gao, Shuchao Henry (2009) *China's strategy for free trade agreements: Political battle in the name of trade*. Research Collection SMU School of Law – Paper 971. http://ink.library.smu.edu.sg/sol_research/971, accessed 29 May 2012.

Garnaut, Ross, Frank Jotzo and Stephen Howes (2008) China's rapid emissions growth and global climate change policy, in Ligang Song and Wing Thye Woo (eds), *China's dilemma: Economic growth, the environment and climate change*, pp. 170–89 (Canberra: ANU Press).

Gathii, James Thuo (2008) Third world approaches to international economic governance, in Richard Falk, Jacqueline Stevens and Balakrishnan Rajagopal (eds), *International law and the third world: Reshaping justice*, pp. 255–68 (Oxford: Routledge-Cavendish).

Gu, Bin (2011) Mineral export restraints and sustainable development – Are rare earths testing the WTO's loopholes? *Journal of International Economic Law* 14(4), pp. 765–805.

Guo, Yingjie (2008) Domestic openness in post-WTO China: Central and local perspectives. *Journal of Contemporary China* 17(55), pp. 339–59.

Heymann, Monika (2008) International law and the settlement of investment disputes relating to China. *Journal of International Economic Law* 11, pp. 507–26.

Houser, Trevor and Roy Levy (2008) Energy security and China's UN diplomacy. *China Security* 4(3), pp. 63–73.

Hsiung, James (2007) Sea power, Law of the Sea, and a Sino-Japanese East China Sea "resource war", in James Hsiung (ed.), *China and Japan at odds: Deciphering the perpetual conflict*, pp. 133–54 (Basingstoke: Palgrave Macmillan).

Jia, Bing Bing (2010) A synthesis of the notion of sovereignty and the ideal of the rule of law: Reflections on the contemporary Chinese approach to international law. *German Yearbook of International Law* 53, pp. 11–64.

Kent, Ann (2001) States monitoring states: The United States, Australia, and China's human rights, 1990–2001. *Human Rights Quarterly* 23, pp. 583–624.

Kent, Ann (2006) Compliance v cooperation: China and international law. *Australian International Law Journal* 13, pp. 19–32.

Kent, Ann (2007) *Beyond compliance: China, international organisations and global security* (Stanford: Stanford University Press).

Kotecki, Stephanie (2008) The human rights costs of China's arms sales to Sudan – A violation of international law on two fronts. *Pacific Rim Law & Policy Journal* 17(1), pp. 209–35.

Lin, Li-Wen (2010) Corporate social responsibility in China: Window dressing or structural change? *Berkeley Journal of International Law* 28, pp. 64–100.

Ma, Xinmin (2007) Statement on the ILC's work on shared natural resources and responsibility of international organisations. *Chinese Journal of International Law* 6(3), pp. 769–72.

Ma, Xinmin (2008) Statement on "prevention of transboundary harm from hazardous activities and allocation of loss in the case of such harm", 2007. *Chinese Journal of International Law* 7(2), pp. 567–69.

Manjiao, Chi (2011) The unhelpfulness of treaty law in solving the Sino-Japan sovereign dispute over the Diaoyu islands. *East Asia Law Review* 6, pp. 164–89.

Marsden, Simon (2010) Developing approaches to trans-boundary environmental impact assessment in China: Cooperation through the Greater Tumen initiative and in the Pearl River Delta region. *Chinese Journal of International Law* 9(2), pp. 393–414.

Meidan, Michal (2010) China's emissions reduction policy: Problems and prospects, in Antonio Marquina (ed.), *Global warming and climate change: Prospects and policies in Asia and Europe*, pp. 307–21 (Basingstoke: Palgrave Macmillan).

Montevideo Convention on the Rights and Duties of States (adopted 26 December 1933, entered into force 26 December 1934, 166 LNTS 19).

Mushkat, Roda (2009) Implementing environmental law in transitional settings: The Chinese experience. *California Interdisciplinary Law Journal* 18, pp. 45–94.

Mushkat, Roda (2010) Compliance with international environmental regimes: Chinese lessons. *William and Mary Environmental Law and Policy Review* 34, pp. 493–542.

Mushkat, Roda (2011) China's compliance with international law: What has been learned and the gaps remaining. *Pacific Rim Law & Policy Journal* 20(1), pp. 41–69.

Posner, Eric and John Yoo (2006) International law and the rise of China. *Chicago Journal of International Law* 7(1), pp. 1–15.

Qi, Dahai (2008) State immunity, China and its shifting position. *Chinese Journal of International Law* 7(2), pp. 307–37.

Ramos-Mrosovsky, Carlos (2008) International law's unhelpful role in the Senkaku Islands. *University of Pennsylvania Journal of International Law* 29, pp. 903–48.

Rosecrance, Richard (2006) Power and international relations: The rise of China and its effects. *International Studies Perspectives* 7(1), pp. 31–35.

Ruggie, John (2011) Business and human rights: Together at last? A conversation with John Ruggie. *The Fletcher Forum of World Affairs* 35(2), pp. 117–22.

Schoenbaum, Thomas (2008) *Peace in Northeast Asia: Resolving Japan's territorial disputes with China, Korea and the Russian Federation* (Cheltenham: Edward Elgar).

Shaw, Han-yi (2008) Revisiting the Diaoyutai/Senkaku Islands dispute: Examining legal claims and new historical evidence under international law and the traditional East Asian world order. *Chinese (Taiwan) Year Book of International Law and Affairs* 26, p. 95.

Shee, Poon Kim (2002) The political economy of China-Myanmar relations: Strategic and economic dimensions. *Ritsumeikan Annual Review of International Studies,* pp. 33–53.

Shen, Jianming (2002) China's sovereignty over the South China Sea islands: A historical perspective. *Chinese Journal of International Law* 1(1), pp. 94–157.

Suganuma, Unryu (2007) The Diaoyu/Senkaku Islands: A hotbed for a hot war?, in James Hsiung (ed.), *China and Japan at odds: Deciphering the perpetual conflict*, pp. 155–72 (Basingstoke: Palgrave Macmillan).

United Nations General Assembly Resolution (22 April 2005) Statement by Ambassador Zhang Yishan on Cluster II (Freedom from fear) of the Secretary-General's Report, informal consultations.
United Nations General Assembly Resolution 1803 (XVII) (14 December 1962).
United Nations General Assembly Resolution 3281 (XXIX) (12 December 1974).
United Nations (2011) *Guiding Principles on Business and Human Rights*, 21 March. UN Doc. A/HRC/17/31.
United States—Definitive Antidumping and Countervailing Duties on Certain Products from China (2011) WTO Appellate Body Report, WT/DS379/AB/R, 11 March.
Van Ness, Peter (1992) Australia's human rights delegation to China, 1991: A case study, in Ian Russel, Peter Van Ness and Beng-Huat Chua (eds), *Australia's human rights diplomacy*, pp. 49–85 (Canberra: ANU).
Wang, Guangya (2005) Statement by Ambassador at Informal Meeting of the General Assembly on the Draft Outcome Document of the September Summit, 21 June.
Wan, Ming (2007) Human rights lawmaking in China: Domestic politics, international law, and international politics. *Human Rights Quarterly* 29(3), pp. 727–53.
Wang, Zonglai and Bin Hu (2010) China's reform and opening-up and international law. *Chinese Journal of International Law* 9(1), pp. 193–203.
World Trade Organisation (WTO) Dispute settlement: Disputes by country. http://www.wto.org/english/tratop_e/dispu_e/dispu_by_country_e.htm, accessed 29 May 2012.
Wu, Xiaohui (2011) No longer outside, not yet equal: Rethinking China's membership in the World Trade Organisation. *Chinese Journal of International Law* 10(2), pp. 227–70.
Wu, Hui and Dan Zhang (2010) Territorial issues on the East China Sea: A Chinese position. *Journal of East Asia and International Law* 3, pp. 137–49.
Xue, Hanqin (2005) China's open policy and international law. *Chinese Journal of International Law* 4(1), pp. 133–39.
Xue, Hanqin (2007) Chinese observations on international law. *Chinese Journal of International Law* 6(1), pp. 83–93.
Xue, Hanqin (2011) Chinese contemporary perspectives on international law: History, culture and international law. *Recueil Des Cours* 355, pp. 47–233.
Zhang, Yishan (2005) Statement by Ambassador at informal consultations on Cluster I: "Freedom from Want" of the UN Secretary-General's Report, 25 April.
Zou, Keyuan (2005) *China's marine legal system and the Law of the Sea* (Leiden: Martinus Nijhoff).
Zweig, David and Jianhai Bi (2005) China's global hunt for energy. *Foreign Policy* 84(5), pp. 25–38.

Securing Fish for the Nation: Food Security and Governmentality in Japan

KATE BARCLAY

University of Technology Sydney

CHARLOTTE EPSTEIN

The University of Sydney

Abstract: *Concerns about supplies of food have been a feature of Japanese politics since Japan started modernising in the second half of the 1800s. It has remained a prominent political issue even after Japan cemented its status as a wealthy country in the 1980s, with the Japanese Government continuing to protect domestic food production from international competition. Protectionism is a curious policy for a country so dependent on world trade, including for food. Protectionist practices have led to entrenched interests in some sections of government and industry. Protectionist ideas are used in nationalist arguments against food imports. The protection of domestic food production, however, resonates positively well beyond the groups that benefit economically from protection and those that indulge in chauvinist notions about the dangers of "foreign" food. The issue, therefore, is broader than interest-group capture or xenophobia. We find it is deeply embedded in Japanese policies relating to food domestically and internationally, and goes beyond government policy as such, involving ways of thinking about protection of national culture, and social and environmental responsibility. Michel Foucault's notion of governmentality helps to explain this approach to food security, accounting for the balancing act between free trade and protection as well as the pervasiveness of this rationality beyond government as such.*

Concerns about supplies of food have long been a feature of Japanese politics. The need to secure access to food, along with other resources, was one of the factors behind Japan's colonial expansion in the first half of the twentieth century and military aggression leading to the Pacific War. After the war and the loss of its colonial sources of food Japan's approach to food security has involved a heavy reliance on world trade for food imports and government support for domestic food industries. The justification for this has been that it was needed to ensure some measure of protection from problems in world food markets, although clearly domestic political considerations were also at play. The Liberal Democratic Party, in power continuously from the 1950s to the 1990s, relied on support from rural electorates in exchange for a range of economic benefits including protection of food production. Since the 1980s Japan has faced opprobrium from food exporting countries objecting to government protection of domestic production, notably for the iconic foodstuff of rice, but also for beef and fruit.

Japan's food security policies have also resulted in international disputes in fisheries sectors. For example, food security is one of the ways in which Japan seeks to legitimise its whaling practices (see Epstein, 2008, pp. 231–36).[2] Japan has also been at the forefront of an on-going disagreement in the Doha round of trade talks over the legitimacy of fisheries subsidies in the context of a global free trade regime. A group of World Trade Organization (WTO) member countries calling themselves the Friends of Fish[3] argue that subsidies to the fisheries sector, making up as much as 20 per cent of revenues or USD25 billion annually, have led to overfishing (Sumaila et al., 2010; WTO, c.2010) and thus should be removed for both free trade and environmental protection reasons. Resisting the push to completely disallow fisheries subsidies are lower income countries needing to encourage fisheries development and also three big fishing countries that make up the bulk of the global total of fisheries subsidies – Taiwan, South Korea and Japan. These countries dispute the connection between fisheries subsidies and overfishing (arguing that inadequate resource management is the problem). Japan for its part asserts that fisheries subsidies are necessary for its food security.

At first glance it seems curious that a country as wealthy as Japan should be concerned about food security at all. Food security first emerged as a problem of global governance in the wake of World War II and the destruction of the food supplies it had wrought; yielding the creation of the Food and Agriculture Organization in 1945. Today, however, food security is generally presumed to be a concern of developing countries. By the definition of the 1996 World Food Summit, food security exists when "all people at all times, have physical and economic access to sufficient safe and nutritious food to meet their dietary needs and food preferences for a healthy and active life" (Pinstrup-Andersen, 2009). The global free-trade regime that was shaped over the decades following World War II was seen by its advocates as the solution to the problem of access. By this logic unfettered trade enables the free flow of foods from producers to consumers, without having to produce all foods domestically. Japan's competitive advantage in the world economy, founded on high value-added industries, should place it in a good position to this end. Moreover, as an advanced industrialised economy with few primary resources, Japan has been highly dependent on trade both for export of its manufactures and import of raw materials, particularly in food, whereby Japan is the world's largest net importer of food (WTO, 2009b). Consequently, Japan has been a strong supporter of international free trade regimes, notably the WTO. For example, Japan strongly opposed the erection of trade barriers in the form of rice export bans by Southeast Asian countries to shore up domestic food supplies in the 2008 food crisis (WTO, 2009a, p. 10; MoFA, 2009).

Yet Japan parts ways with the WTO on its understanding of food security. In Japan, discussions of food security are always about food self-sufficiency. In defining food security the WTO starts from the World Food Summit definition, but then adds a sentence: "'Food security' and 'self-sufficiency' are not the same and a key debate is whether policies aiming for self-sufficiency help or hinder food security" (WTO, n.d.). Differences in definitions of food security reflect different policy positions. Japan laid a strong claim to framing the food security question when, in the lead-up to the World Food Summit held in the FAO's headquarters in Rome, it hosted, in collaboration with the FAO, the International Conference on the Sustainable Contribution of Fisheries to Food Security. With more than 500 participants, the Conference yielded the Kyoto Declaration and Plan of Action on the Sustainable Contribution of Fisheries to Food Security (Fisheries Agency, 1995), which gave prominence to the understanding of food security as self-sufficiency, by means of the concept of sustainability.[4]

How does a traditional proponent and beneficiary of free trade regimes, which is highly dependent on world trade, arrive at a position where trade protection is defensible policy? In this paper we show how Michel Foucault's concept of governmentality offers a conceptual framework for accounting for this singular combination of being open to world markets while also protecting domestic industries that characterises Japan's food security policies, particularly in the area of fisheries. Foucault (2007) coined the concept in his 1977–78 Lecture at the College de France as a counterpart to "sovereignty", in order to capture the new forms of state-market relationships in which the modern state was increasingly involved. "Governmentality" is a broadly writ concept for the wide range of modes of thought and practices permeating late capitalist societies that are primarily geared towards enhancing the population's productive capacities, as a key resource in the capitalist production system.

In the first part of the paper we theorise food security within the terms of Michel Foucault's conceptual framework. We illustrate how, more than similar policies designed to support national competitiveness on the world scale, food security represents the archetypal governmental objective of the modern state. In the second part of the paper we trace a history of food security in Japan in the fisheries sector, particularly the role of state support of domestic production, using governmentality as an explanatory framework. We find there are entrenched interests in government and industry for government support for Japanese fisheries and consider whether the politics of these interest groups is a sufficient explanation for Japan's approach to food security. We find that the Japanese approach to food security extends beyond the national government, being visible in public opinion on issues as varied as the protection of national culture, rural social stability and coastal environmental stewardship. We also find the approach articulated in Japan's international relations beyond its food trade policy, in its official development assistance (ODA) and other forms of diplomacy.

Between Sovereignty and Governmentality: Framing Food Security

Michel Foucault coined the concept of governmentality in his 1977–78 Lecture at the College de France *Security, Territory, Population* (2007). He takes as a starting point for his historical enquiry the state's obligation to secure a territory for its population. The provision of security, Foucault finds, is the founding obligation of sovereignty, still defined by way of its two attributes, possession of a territory and a population. This security obligation set the terms within which "sovereignty" has been theorised by Machiavelli onwards since its emergence in the Renaissance. It was to break beyond the limits of "sovereignty" thus conceived that Foucault coined the concept of governmentality, finding that there was much more to the contemporary neoliberal state taking shape in the late twentieth century (the backdrop against which Foucault was theorising) than was encapsulated in theories of sovereignty. He thus offered the concept of "governmentality" as counterpart to "sovereignty" to bring to light the expanding range of domains and new forms of power in which the contemporary state had become increasingly involved, and that were largely obscured by a focus on sovereignty alone.

It is noteworthy that in developing his concept Foucault identified markets for food as a key site for the development of the modern state and liberal capitalism (Foucault, 2007, esp. pp. 51–71, pp. 443–55). Revolutions in Europe made rulers aware of the political importance of riots caused by high prices for grain. Moreover, during the mercantilist era food markets were seen as important because low grain prices enabled low wages, facilitating the export of manufactures and imports of gold. Then both mercantilism and attempts to control supplies of grain were discredited by political economists who found that markets operated according to underlying principles that tended to set food prices at the right level for increasing national wealth as long as state attempts to control markets did not block the operation of these principles. The freedom of trade between countries thus entered government rationalities. Another important realisation was that low prices were not necessarily in the national interest, because alongside urban consumers, rural producers also needed to be kept content and productive. Letting the markets unfold, it was soon realised, shifted the price to the optimum level for the good of both urban and rural populations. Governments therefore reoriented their policies from controlling markets to enabling them.

Governmentality thus captures a trajectory of the modern state, which, as European territorial boundaries were increasingly settled, began to reach beyond the traditional remit of sovereignty (the securing of a territory and a population), and into the very fabric of society itself. The term captures the state's progressive penetration into the economy at large, and the new forms of knowledge that developed accordingly (notably political economy). Hence "governmentality" evokes a historical process characteristic of the modern era that has yielded the contemporary neoliberal state, a state intimately bound up with markets. Michel Foucault thus posed sovereignty and governmentality as the conceptual poles framing the range of the neoliberal state's activities.

The advantage of Foucault's conceptualisation lies in the way he continues to foreground the centrality of sovereignty and the associated themes of security and territory in the operation of the neoliberal governmental state.[5] There is a large body of work devoted to the topic of openness in world trade. Ever since it has been recognised, in the wake of Karl Polanyi's (1944) *The Great Transformation*, that economic activity is always socially embedded, International Political Economy (IPE) scholarship has analysed the multiple and varied accommodations that states have made with markets in different contexts; in the embedded liberalism strand (Ruggie, 1982), the literature on the competition state (Cerny, 1997), economic nationalism (Clift and Woll, 2012), the cultural political economy approach (Jessop and Oosterlynck, 2008), regulation theory (Boyer, 2005), the varieties of capitalism approach (Hall and Soskice, 2001), and, in the case of Japan specifically, the developmentalist state literature (Johnson, 1995). We find, however, that governmentality lends itself to theorising food security more aptly than other approaches that have examined how states negotiate the contradictory pulls between opening their economies to and protecting them from world markets. Food security mobilises both sovereignty and governmentality. Towards the sovereignty pole, it accounts for the framing of the issue area in terms of security, because it illuminates a state that continues to cater to its traditional sovereignty obligations to secure a population against, in this case, insufficient food supplies within the territory. Towards the governmentality pole, food as an area of governmental intervention constitutes both a productive sector of the economy in its own right, and one that hails more fundamentally the chief purpose of governmentality, enhancing the population's productive capacities.

More than other areas of state intervention, food security policies go to the heart of the governmental state and the new form of politics it set in motion, *biopolitics* – that is, where the entire orientation of state activity (including policymaking) is dedicated to enabling life and its productive capacities. Food security exemplifies governmentality as a mode of rule that, to use Aihwa Ong's words "harnesses and extracts life forces" according to "principles of discipline, efficiency and competitiveness" (Ong, 2006, p. 4, p. 13) insofar as its target enables life itself. Being central to the population's vitality, food security is closely bound up with *biopower*, the power to enhance life, the specific modality of power associated with governmentality.[6]

Governmentality as an approach to the study of society has also significantly broadened the question of "who governs". By encompassing everything from the dispositions and outlooks of the governed to the regulatory and institutional mechanisms through which they are governed, governmentality calls into question public-private and state-market binaries. As Tim Luke underlines:

> The notion of governmentality invites social theorists ... not to reduce the complicated ensemble of modernising developments to the actions of "the state" [but instead] to investigate the "governmentalisation" of the economy and society (Luke, 1995, p. 27; see also Ferguson, 1994).

Governmentality draws out a series of continua rather than breaks between state and non-state practices, public and private institutions. This part of the concept is particularly useful in explaining Japanese approaches to food security in fisheries.

The Role of Fish in the Modernisation of Japan: The Rise of Japanese Fisheries

As a densely populated island country, Japan has historically turned to the sea as a key source of protein. The eating of four-legged animals was banned by the government from 675 CE, largely for religious reasons. Although the ban was lifted in the 1860s, cultural patterns and the economics of raising animals for food meant consumption of animals other than chickens or fish remained low for another century. Japanese fisheries were thus an integral part of plans for the modernisation and industrialisation of Japan, for eliminating the famines that occurred in some areas and for coping with problems caused by accompanying urbanisation, particularly preventing the impoverishment of rural communities. In the late 1800s, the Japanese Government responded to this problem with a policy of redistribution to rural areas of wealth generated by industrialisation in urban areas (Francks, 2006), a policy that has continued ever since (Mulgan, 2000). In this context domestic fisheries were important for building food supplies, and for providing income for rural communities. As urbanisation, industrialisation and population growth increased the demand for food, however, domestic fisheries were soon insufficient.

In the decades preceding World War II various countries, including Japan, saw autarchy in resources as necessary for their security. The Japanese Government was not content to leave the supply of resources up to world markets, but actively guided and supported its resource industries to promote production and supply. Japanese fishing companies were encouraged by government to innovate technologically to increase production, and were also encouraged to expand their operations outside Japan ("distant water" fisheries). Japanese fisheries companies moved into fishing grounds in neighbouring countries and eventually down into Southeast Asia and the Pacific Islands region as part of the strategy for supplying the empire with resources.[7] The legislative framework was the Meiji-era Fisheries Promotion Act (1897). Early initiatives included the financial support for technological developments in ship-building, such as installing engines (1903), refrigeration equipment (1907) and radios (1918) (Fujinami, 1987).

Japanese Government support of industries has mostly taken place through a layer of affiliated organisations (*gaikaku dantai*) that sit between ministries and the industries they govern. Bureaucrats, who are very powerful relative to politicians in the Japanese Government (Mulgan, 2002), direct government intervention in the economy, including through "administrative guidance" (Amyx, 2003). Administrative guidance and the funnelling of public funds to industry are conducted through the *gaikaku dantai* organisations. These may be special, public and chartered corporations that engage in public works or financial services for particular sectors, or may indeed be industry associations. The organisations are directly managed and subsidised by ministries in relation to the sectors under their portfolio. There are thousands of these types of organisations in Japan and vast amounts of public money flow through them (Carpenter, 2003).

In fisheries the key *gaikaku dantai* have included bodies such as the Japan Fisheries Association (established 1882) and the Fisheries Cooperative Associations (their forerunner was given fishing rights to allocate to industry under the Fisheries Law in 1901). Rather than being managed and run by industry people, in the *gaikaku dantai* system industry organisations were established by government, have been funded by government, and their senior positions have been filled by former bureaucrats, politicians and members of the imperial family. The functions of these organisations have included activities that might in other countries be done by government, such as regulation of fisheries.[8] They have also included functions that might normally be done by the private sector, such as human resources and payroll services for fishing companies, and acting as agents for distant water vessels in foreign ports, and as such they constitute a form of subsidy. Other kinds of subsidised services provided to fisheries through *gaikaku dantai* included pension funds, insurance against vessel damage/loss and poor catches, and various kinds of financial services from basic banking to high risk venture capital.[9]

The fisheries *gaikaku dantai* system continued to develop after Japan's defeat in 1945. When the Allied forces occupied Japan they were confronted with an economy devastated by war. There were severe food shortages for several years. Not only was Japan's domestic food production and distribution system in disarray, the colonial food production Japan had been relying on for several decades was gone. Considering the limited supply of arable land, fisheries were key to fulfilling Japan's protein needs. Japan's fishing fleet had played a strategic role in Japan's military expansion, thus in the initial postwar period the Allied forces banned Japanese fishing interests from venturing outside Japan. But as it became clear that Japanese coastal fisheries production would not produce enough to meet the country's dietary needs, and as Japan's position in the Cold War became more important, the US-led Allied occupation administration reversed this policy. Indeed, the US came to have the same view as the prewar Japanese leaders; that self-sufficiency was important for Japan's food security, and that government support for the industry, including support for its operation outside Japanese waters, was necessary to achieve this goal (Smith, 2008). General MacArthur thus encouraged the re-establishment of a national whaling industry, funded with US capital (Epstein, 2008). When the Allied occupation ended in 1952 Japanese leaders continued policies of administrative guidance and support for the sector, along similar lines to the prewar period. With the recovery of the Japanese economy and government encouragement, fisheries production grew to a peak in 1964 when Japanese production exceeded domestic seafood consumption (Hayes, 2010).

As well as domestic production and Japanese distant water fisheries, the Japanese Government has also sought to achieve food security through diversifying supplies from overseas. Since the postwar era much of this project has been conducted through the provision of fisheries aid and technical assistance to developing countries (Tarte, 1998). One main source of government support for developing overseas sources of supply has been the Overseas Fishery Cooperation Foundation (OFCF), a *gaikaku dantai* run by the Fisheries Agency of the Ministry of Agriculture, Forestry and Fisheries (MAFF). The OFCF funds technical experts to go out and work in seafood production outside Japan, including in developing countries, and sponsors people working in fisheries and fisheries management to come to Japan for training. It also offers low interest loans for the development of overseas seafood industries and in this endeavour has supported Japanese companies to develop food-producing ventures. For example, seafood giant Maruha was involved from the early 1970s to 2000 in a skipjack fishing and processing venture Solomon Taiyo in the Solomon Islands. This company was majority owned by the Solomon Islands Government, but was also a part of Maruha's overseas operations branch, and was an important source of product for Maruha in fulfilling its trading contracts with UK buyers such as Sainsbury's. It also constituted one of the main imported sources of *katsuobushi*[10] for the Japanese market. OFCF provided several millions of dollars (US) worth of low interest loans to Maruha for developing the Solomon Taiyo factory in the early 1990s (Barclay, 2008). OFCF assistance thus encouraged overseas production to supply the Japanese market, both by foreign companies and by Japanese seafood companies operating transnationally.

From the 1970s Japanese fisheries encountered problems to do with oil price rises, the beginning of territorial enclosure of the seas through the declaration of 200 nautical mile Exclusive Economic Zones under the United Nations Convention on the Law of the Sea, and competitiveness problems against cheaper production locations (Barclay, forthcoming). Various forms of subsidy were put in place to try to halt the decline of Japanese fisheries (Campbell and Nicholl, 1994; Fujinami, 1987; Bergin and Haward, 1996; Kagoshima Prefecture Skipjack and Tuna Fisheries Association, 2000). Nevertheless Japanese fisheries production has continued to decline, especially since the 1990s. Compounding the lack of competitiveness with imports from cheaper production locations, as Japan's wealth has grown, the Japanese workforce has chosen not to enter physically demanding, low-status fields such as fishing. In the absence of an immigrant community willing to step into the breach there is a severe workforce shortage (Hori, 1996; Government of Japan, c.2003; Hayes, 2010). From 1972–88 Japan was the world's largest fisheries producer by volume. Since the 1990s China and Peru have exceeded Japanese fisheries output (Government of Japan, c.2003). Japanese production continues to fall and many see Japanese fisheries as being in a "terminal decline" (Biggs, Matsuyama and Balfour, 2011).

With the reduction in domestic production, Japan has come to rely more heavily on seafood imports. In 2002 Japan was the leading importer of fishery commodities, importing to the value of USD13.6 billion (thousand million). In that year the next biggest importer was the USA with around USD10 billion and the next biggest was Spain with around USD3.8 billion (WTO, 2005). Japan's fishery product imports have decreased slightly in volume since then, and in 2005 China overtook Japan as the world's largest importer of fishery products (FAO Fishstat data, cited in Government of Japan, 2010, p. 11).

In the 2010 White Paper on Fisheries, increasing Japan's self-sufficiency in seafood is listed as one of the top policies, and concerns about supplies of fishery products are raised several times throughout the document (Government of Japan, 2010). The Japanese Government enacted The Basic Law of *Shokuiku* (Food Education) in 2005. The rationales for the law and the policies flowing from it targeting agricultural and fisheries production include "over-dependency on food from abroad" (MAFF, c.2005). Masayuki Komatsu, a prominent fisheries commentator and former senior bureaucrat in MAFF who has led Japan's country delegations to the International Whaling Commission and the international bodies that manage tuna resources globally, says:

> Japan cannot continue simply relying on imported food. Can we afford as a country to be dependent on others, such as the United States or Australia, for our basic foods? Will we always have enough precious dollars to import what we need? It is the answer to these questions that should tell you why I firmly believe that we need to become more self-sufficient for reasons of our national health and at the most basic level, to guarantee the supply of food to our people (Smith, 2004).

These concerns with both self-sufficiency and production have been compounded by the recent effects of the tsunami and radiation contamination from the Fukushima power plant in 2011. The Pacific coast north of Tokyo is the location of many of Japan's fisheries, such as the port of Kesennuma, one of the ten largest fishing ports in Japan. Vessels, ports, market areas and processing plants were all ravaged by the tsunami. Kesennuma's fish market was planning to partially re-open in June 2011 but according to the local Fisheries Cooperative Association and local politicians, central government assistance is necessary for long-term reconstruction, and that reconstruction is necessary to prevent a large-scale outflux of workers (Biggs, Matsuyama and Balfour, 2011). Fisheries *gaikaku dantai* have been assisting fishing ports such as Kesennuma ever since the disaster struck, at first with food and medical supplies, but they also aim to continue to support the industry to rebuild (OPRT, 2011). This devastation has reinforced the Japanese Government's commitment to its long-term policy of supporting rural food-producing communities. It has already proposed the start of a series of aid packages for the regions most affected by the earthquake, and the fishing industry will be one of the beneficiaries (Biggs, Matsuyama and Balfour, 2011). This catastrophic natural disaster brings out in the starkest fashion that, in addition to the organisational dimension of the governmental state regarding markets, the modern state's biopolitical imperative is to sustain life.

Whether the policies Japan has pursued in the fisheries sector have successfully secured the maximum amount of fish for the nation, or whether this might have been better achieved through a different policy mix (including more openness to international trade), is an important empirical question, but it is beyond the scope of this paper. This paper's concern is with the underlying rationale with which Japan has framed, justified and garnered support for its food security policies.

Government Support for Domestic Fisheries Production: Just Institutional Capture?

The modern Japanese state is well known for intervening in the economy. The first wave of literature accounting for the Japanese state's extensive involvement in markets emphasised the role of bureaucrats motivated by national interest (Johnson, 1995). A more recent strand of analysis, however, has pointed to evidence that bureaucrats have been motivated by personal gain or the gain of their group within government, and for the economic development of some groups ahead of others rather than for the nation as a whole (Mulgan, 2005; Hein, 1994). It should be asked, therefore, whether the Japanese Government position on support for domestic fish production in the name of food security is in fact a manifestation of Japanese governmentality for the national good, or is instead a case of institutional capture of the national government agenda on food security by those with vested interests in continued government support for fisheries. Institutional capture usually refers to the capture of government institutions by an interest group outside government. As we have seen, however, in Japan interests come to be aligned between industries and the parts of government with jurisdiction over them. In this case, therefore, the interest group is made up of both government and industry actors.

The *gaikaku dantai* system outlined above clearly contains much scope for pursuing personal and group interests in the name of the national interest. Most fisheries are based in rural areas, and it has been well documented that the Liberal Democratic Party (which was in power continuously from 1955–93 and 1994–2009) made use of public spending in rural areas to shore up its electoral base (Mulgan, 2000; DeWit and Yamazaki, 2004).[11] Aurelia George Mulgan (2005) has argued convincingly that bureaucrats overseeing food production under MAFF have acted to protect their sphere of influence in industry and the size of budget they control through the *gaikaku dantai.* They thus engage in intervention in the economy specifically to maximise and defend the order that benefits them collectively (as a ministry, and as agencies within the ministry) and individually. A study in a similar vein explains Japan's whaling policy by Fisheries Agency bureaucrats building and guarding their "turf" in terms of sphere of influence and budgetary allocation for the scientific whaling program (Ishii and Okubo, 2007). MAFF officials have been involved in several scandals about the misuse of government money and taxpayer-funded initiatives for personal gain (Amyx and Drysdale, 2003; Mulgan, 2003, pp. 171–74).

As might be expected, industry interests are closely tied to the status quo of government support for fisheries. For example, many of the distant water tuna fishing companies in Japan receive heavily subsidised loans through the *gaikaku dantai*, and thus feel unable to speak out against policies with which they disagree (Hori, 1996). Also, industry groups are managed by former bureaucrats (Barclay and Koh, 2008). Businesses that disagree with their industry groups fear retribution both from the ministry (for example, withholding of licences) and from the group (for example, a boycott by other group members) (Carpenter, 2003). The *gaikaku dantai* system has thus served to very strongly entrench particular sectoral interests in the established system of government support of food producing industries.

Institutional capture is indeed at work here. It is undeniable that ideas on food security have been used by sectoral interests to justify policies that benefit themselves. We argue, however, that this is only a partial explanation for the Japanese position on food security. The population at large generally agrees with MAFF's conception of food security. A government opinion poll found that, in 2010, 86 per cent of people surveyed were anxious about being able to import food in the future and 91 per cent felt food self-sufficiency should be increased (MAFF, 2010).[12] This perception is reflected in consumer movements from across the political spectrum. Members of non-government organisation Consumers Japan (Shōdanren), whose Secretary General Ms Hisa Anan openly advocates abandoning the lethal scientific whaling program, feel the prevailing level of food self-sufficiency of around 40 per cent is too low, and that a rate of 80 per cent or higher would be more appropriate.[13] Food self-sufficiency is a central concern of the Consumers Union of Japan (Nishōren), a participant in the World Social Forum.[14] From a more nationalist perspective the consumer group Women's Forum for Fish (WFF) was started by pro-whaling activist Yuriko Shiraishi in 1993 because she felt Japan was vulnerable due to its low self-sufficiency in seafood (Shiraishi, 1999). While MAFF communication strategies to promote its protectionist policies have no doubt played a role in this broad consensus, there are also other reasons so many Japanese people feel as they do about food security.

Dependence on imports is a common media discourse in Japan. It both expresses food security concerns among the public and helps to reproduce them (Bestor, 1999, p. 167). The rise of the governmental state and the history of food security in Japan during the twentieth century help to explain the general consensus that self-sufficiency in food is an important goal for government, and that consumers should side with producers in a joint effort working towards this goal. This has meant that consumers have been willing to pay much higher than the world price for rice (MAFF, 2010), and are happy for their taxes to be used to support Japanese producers (Maclachlan, 2004).

Japanese whaling policy also illustrates the limits of the institutional capture explanation. A narrow focus on bureaucratic interests underplays the extent of the cost of pursuing national whaling policies, both financially and in terms of damage to Japan's international reputation. The persistent pursuit of whaling policies notwithstanding such costs is instead the expression of a much wider consensus, both amongst the government and the population over a long period of time, explicitly in terms of "food security".[15]

Entwined with the concern about access to food and doubt that world markets can be relied on to deliver this is a concern about cultural preferences in food, and whether globalisation will destroy local food cultures. Food is a prominent part of culture at all levels in Japanese society. It is used as a marker of cultural identity, and government support for fisheries is widely seen as legitimate because of the need to preserve cultural heritage in certain food production areas (Bestor, 2004; 1999; Epstein, 2003). "Fish food culture" – the art of fishing, preparing and consuming fish – is seen by many as an essential part of being Japanese. One way this identity is performed is through representing Japan as having a fish food culture as opposed to the "meat food culture" of the West (Bestor, 1999; Hirata, 2004). The sense that the West opposes Japanese fish food culture is bolstered by representations of Japan "guzzling" excessive amounts of endangered species of fish (Renton, 2005), and recurrent depictions in the widespread and well-entrenched anti-whaling discourse of Japanese practices of whaling and dolphin fishing as barbaric (see Epstein, 2008).[16]

In the wake of the rise of the global sustainable development discourse in the 1990s, the Japanese Government's support for fisheries production is increasingly bound up with notions of social and environmental sustainability in rural coastal areas. Foucauldian scholars have analysed extensively the ways in which concepts of sustainability and environmental issues more broadly directly mobilise the biopolitical functions of the governmental state, in that they go to the heart of its life-enabling imperative[17] – that is, in terms of both the well-being of its population and everything that sustains it, including biological and ecosystemic capacity (Darier, 1999; Luke, 1999; 1995). Even though MAFF policies (such as allowing the Japanese fleet to over-catch southern bluefin tuna for many years) have exacerbated overfishing, and MAFF use of the term sustainability may be viewed cynically as an attempt to legitimise practices protecting sectoral and/or national interests, the way MAFF frames this taps into a deeper governmentality rationale. This is visible in the way MAFF addresses the issue of on-going support for fishing and other food-producing communities. For example, the driving concept of the 2010 White Paper on Fisheries is a holistic notion of sustainability, including environmental conservation as well as social and cultural reproduction. A brief exposition of the idea is followed by a diagram "compiled by the Fisheries Agency based on a report by the Science Council of Japan" showing a piece of land from mountains in the background down to coast in the foreground, with farming land, residential areas, factories and fisheries (Government of Japan, 2010, p. 7). The landscape is dotted with labels showing the functions of human activities as well as those of the natural landscape. For example, tidelands and seaweed beds are labelled "water purification" and "eco-system services". A fishing vessel is labelled "providing opportunities for exchanges", "inheriting traditional culture, including traditional fishing methods", "fishing to complement nitrogen and phosphate cycles" and "marine salvage, border surveillance, disaster relief, marine environmental monitoring". This framing of fishing communities puts fishing activities, as well as the other activities in which fishing people engage in their localities, as being important not only in terms of food production for the nation, but also for culinary cultural heritage, the maintenance of rural communities, and conservation of the environment. The support of rural communities is also a central motivation for the continuation of whaling (Epstein, 2008). These arguments as to why Japanese people should want fishing communities to thrive have proven persuasive. A survey conducted by MAFF in 2009 found the public felt the top four roles of fisheries included "supply of food to people" (95.2%), "eco-system conservation" (66.9%), "inheritance of traditional culture, including traditional fishing methods" (55.9%) and "water purification and marine environment conservation" (47.7%) (Government of Japan, 2010). Other arguments about the positive sustainability effects of local production relate to the problem of carbon emissions and the "food miles" in imports (Mashimo, 2009; Hasegawa, c.2010).

This alignment of norms about government support for food production across government and society, a key characteristic of governmentality, has been actively promoted by interested stakeholders. In a report on a MAFF-sponsored symposium on world food security held in 2009, participants (a mix of industry, non-profit and non-government organisations, government and *gaikaku dantai* organisations, academia and the media) agreed that it was important for the Japanese public to "have a sense of crisis on their vulnerability" regarding the food situation (MAFF, 2009). The Fisheries Agency has engaged in a public communication strategy to promote the idea that whaling is in the national interest, at first through a discourse that whaling was important nutritionally (food security). Later, as the economy recovered and the level of consumption of whale meat fell to a tiny proportion of people's daily diet (most Japanese people do not eat whale from one year to the next), the strategy has been to show that the international ban on whaling constitutes cultural imperialism on the part of the Western non-whale-eating powers and is thus an attack on the cultural traditions of Japan. International disputes about whaling have since come to be understood in these terms by the Japanese public (Ishii and Okubo, 2007; Hirata, 2004).

The fact that the Japanese approach to food security extends beyond MAFF, therefore, does not disprove the existence of institutional capture because MAFF has spread these ideas through communication campaigns. Nevertheless, it would be too deterministic to say these campaigns are the only reason the approach is so pervasive. The concern to protect domestic production is credible enough, in the context of Japanese governmentality, to resonate throughout society. Japan is exposed to world food markets, and empirically these markets fail from time to time. Japan has had food shortages within living memory. Rural communities are in crisis, and coastal environments are degraded (in part due to fishing activities). Japanese food culture has been subject to cultural imperialism by some Western anti-whaling commentators. Without denying the play of institutional capture, we therefore argue that more is at stake in Japan's approach to food security, and here the concept of governmentality sheds some useful light. This is relevant for questions of policy reform broached by the literature on institutional capture. If it were just institutional capture then restructuring MAFF and its use of public money through *gaikaku dantai* would be the way to deal with it. If, as we argue, it is governmentality, then the approach permeates more deeply into a range of institutions, and change would require broad-scale shifts in understandings about the role of government in the economy, the reliability of markets in providing public goods, national food culture, and how best to govern rural coastal areas, not only within government but throughout society.

Japan's Approach to Food Security in the International Sphere

The Japanese approach to food security – understood as requiring government support for domestic producers to ensure some measure of self-sufficiency – is deeply entrenched in Japanese governmentality. In the international sphere, however, prevailing governmentalities, sustained by organisations such as the WTO, posit free trade as the best way to secure food supplies. Furthermore, in the international sphere food security is usually understood as a problem for low-income and developing countries rather than wealthy countries such as Japan. In this section we consider the diplomatic dimension of the Japanese governmental state's strategies to promote its approach to food security in the international sphere.

Japan has taken a leadership role in convening international meetings on food security, and thus helped shape the international debate in ways that promote Japanese perspectives on food security. Fisheries are a key source of protein for many of the world's poorest communities. With declining productivity in many of the world's fisheries due to overfishing and other human impacts on the marine environment, fisheries are at the centre of concerns about food security. In addition to the internationally-important issues addressed by Japanese diplomacy, however, the ways some of these issues are framed also promote the Japanese national approach to food security. One is the prominence of the word "sustainable" in the title of the Declaration and Plan of Action on the Sustainable Contribution of Fisheries to Food Security. That fisheries should be pursued sustainably is an unarguable point. However, in Japanese discourse, "sustainability" is explicitly bound up with food self-sufficiency, as we have seen.

The understanding of food production as not just an economic activity but one that must be considered in conjunction with its social, political and environmental outcomes was followed up in the Japanese address to the World Food Summit in 1996 by the then Minister of Agriculture, Forestry and Fisheries, Takao Fujimoto:

> [T]he promotion of domestic production will facilitate the full deployment of the multiple functions of agriculture other than food production … we believe that in order to achieve food security, it is particularly important for each country to produce domestically at least its basic food and to pursue sustainable agriculture and rural development (WFS, 2006).

The Japanese Government has continued to give prominence to the ideal of sustainable production in pronouncements on food security for developing countries. In 2009 Japan hosted an international roundtable on Promoting Responsible International Investment in Agriculture, in conjunction with the FAO and several other prominent intergovernmental organisations. At this meeting food production was described by the Japanese Ministry of Foreign Affairs (MoFA) as something involving "political stability, social cohesion, human security, sustainable food production, household food security, and environmental protection". It was hoped that the workshop would "promote responsible investment in agriculture that will achieve sustainable and inclusive agricultural development" (MoFA, 2009). In the same year, MAFF held a domestic symposium on world food security, in which the Senior Vice-Minister spoke about the importance of "sustainable agriculture and the improvement of the food production in developing countries" (MAFF, 2009).

In addition to speaking up on how food security is defined internationally, Japan has also furthered its approach to food security through its Official Development Assistance (ODA) program. Supporting food production in overseas countries, including developing ones, has been one of the key strategies of Japan's handling of international sources of food supply since the demise of its empire. For example, as a result of Japan's experience of the 1973 US soybean export embargo, Japan sought to diversify its soybean supplies, including through its ODA. One of the areas it focused on was soybean production in the Cerrado region of Brazil, which significantly increased production (MoFA, 2009), giving Japan another stable source of supply for this staple. In his address to the World Food Summit in 1996 Fujimoto noted the contributions Japan makes to food security through its ODA (WFS, 2006). Comparatively Japan has not had a large percentage of GDP devoted to aid, but because the size of its economy has been so large since the 1980s it has in absolute terms been one of the largest aid donors over the past few decades. At the time of the World Food Summit Japan was the largest ODA donor country. Japan has long been a big fisheries aid donor. As well as the Overseas Fishery Cooperation Foundation (OFCF) initiatives noted earlier, which are deployed worldwide to support fisheries production that may supply the Japanese market, the Japanese Government has been a long-time sponsor of small-scale coastal subsistence and livelihood fisheries in low-income and developing countries. In the Pacific Islands region Japan has since the 1970s provided on-going support for coastal fisheries supplying local markets in terms of ice machines, vessels, extension services for fishery techniques, and wharf and market infrastructure (Barclay and Cartwright, 2007; Tarte, 1998).

This is not to say that Japanese policies have always delivered outcomes in line with its vision of sustainability in food security. Increased soybean production in Brazil has been linked to deforestation. Japanese fleets have overfished as much as those of any other country, and Japan continues to be the main market for overfished bluefin tuna stocks. The Japanese position on food security is a complex mix of practices and discourses that include narrow sectoral interests, national self-interest, and mutual interest with humanity at large. These ideas about food security have traction because, however instrumentally they may have been used by some actors, the ideas themselves are credibly in the public interest, and capture the concerns of many people beyond those who cynically benefit from practices carried out in the name of those ideas. In other words, they tap into deeper governmentality rationales underpinning the modern Japanese state.

Japan has thus for several decades indicated through its ODA budget, and the overseas activities of *gaikaku dantai* organisations such as the OFCF, that it takes food production very seriously as an issue for low-income and developing countries. Japan's approach to food security, juggling reliance on world markets with protection of domestic production, is woven throughout various aspects of Japan's international relations. Japan's leadership role both as a major ODA donor in food sectors and as a host of international meetings about food security puts it in a position to promote this dual agenda of making sure food exports are free and functioning smoothly, while also protecting domestic production from global competition.

Conclusion

The Japanese approach to food security goes to the heart of Japanese governmentality. Food production is seen as central to the state's role vis-à-vis the population. The promotion of free international trade while also protecting domestic production is broadly seen as the "right" way to do food policy within government circles and throughout society. Because of its focus on productive capacity the concept of governmentality could also be usefully applied in analyses of energy security, and also for the broader concept of human security. The significance of understanding the Japanese approach to food security in this way is that it highlights how the approach is entrenched through a range of apparently unrelated institutions and areas as different as whaling and tuna fishing. This means a change to the approach could not occur simply through altering policies on trade barriers and subsidies to food producers, or breaking up the entrenched interests in continued protection between government and food producers. Such a change would involve a fundamental rethinking of government obligations to the population regarding food as a whole – as a physical necessity subject to failures in world markets, as the foundation of rural economies, and as a valued part of national culture. It would also require a rethinking of Japanese diplomacy regarding food, including its ODA. For changes in food policy to have a receptive environment, changes would need to occur in popular understandings of the way food production and food security *should* be done.

Acknowledgments

This paper was much improved by helpful comments from Shiro Armstrong, Graeme Smith and Peter Drysdale, as well as two anonymous reviewers.

Notes

1. Our point here is not to engage in whether such a defence of whaling is spurious (which it no doubt is) but rather to draw attention to the extent to which food security has permeated the framing of Japanese fisheries policies.
2. Members of the Friends of Fish group include Argentina, Australia, Chile, Colombia, New Zealand, Norway, Iceland, Pakistan, Peru and the USA.
3. Centrally, in the Kyoto Action Plan the notion of "sustainability" enables the emphasis to be placed on the *production* of food (in this case, fish) rather than on *trade* as a means of securing the provision of foodstuffs. Moreover, "sustainability" is also arguably a red flag with regards to the WTO, given the latter's poor environmental record and its perceived role in forcing through free trade at the expense of national environmental protection initiatives. Lastly, historically tensions between the FAO and the WTO are not new.
4. We use "neoliberal" and "governmental" state interchangeably.
5. In his lecture Foucault (2007) shows how the state's increasing involvement in fuelling and funnelling the accumulation of riches called for new forms of power, where power was no longer exercised as top-down expression of sovereign power (the power to kill) but rather bottom-up, productive power to enable and enhance the population's own productive capacities, such as disci-

pline (which is exercised over the individual) and biopower (which targets the population as a whole).

6. For more expansive analysis of the role of fisheries in Japan's imperial expansion, see Koh and Barclay (2007) and Chen (2009).
7. The Japanese system of fisheries regulation is thus famous for being an effective and efficient form of "co-management" (Makino and Matsuda, 2005).
8. For a more expansive discussion of *gaikaku dantai* in fisheries see Barclay and Koh (2008).
9. *Katsuobushi* – smoke-dried and mould-cured skipjack – is a widely used condiment and flavouring in Japanese cuisine.
10. For a case study of public spending on tuna fisheries being used for political ends, see Barclay and Koh (2008).
11. The source does not reveal the design of the survey, which could have been constructed to elicit such an emphatic result.
12. Interview by Kate Barclay and Sunhui Koh with Shōdanren (Consumers Japan) members Toshiko Kanda, Yasue Itō and Takako Hasuo, Tokyo, May 2003. Since 1998 Japan's self-sufficiency ratio in food calculated on a calorie basis using the Food Balance Sheet method has been around 40 per cent, the lowest among wealthy countries. South Korea is under 50 per cent, Switzerland between 50 and 60 per cent, Australia is over 200 per cent, and France and the USA are both over 120 per cent (Hasegawa, 2010).
13. As evidenced by its website, which features papers such as 'To What Level Could Japan's Food Self-Sufficiency Recover?' (Mashimo, 2009) under the 'Food Security' tab.
14. See Epstein (2008) for an extensive development of this argument.
15. A well-known example of this kind of representation is the award-winning documentary *The Cove*, directed by Louie Psihoyos.
16. Once again we are setting aside the issue of whether this is done most efficiently: even if it is done less efficiently than another mix of policies would do it, it still sustains the life-enhancing goals of governmentality.

References

Amyx, J. (2003) The Ministry of Finance and the Bank of Japan at the crossroads, in J. Amyx and P. Drysdale (eds), *Japanese governance: Beyond Japan Inc*, pp. 55–76 (London: RoutledgeCurzon).

Amyx, J. and P. Drysdale, eds. (2003) *Japanese governance: Beyond Japan Inc* (London: RoutledgeCurzon).

Barclay, Kate and Ian Cartwright (2007) *Capturing wealth from tuna: Cases from the Pacific* (Canberra: Asia Pacific Press).

Barclay, Kate (2008) *A Japanese joint venture in the Pacific: Foreign bodies in tinned tuna* (London: Routledge).

Barclay, Kate and Sunhui Koh (2008) Neo-liberal reforms in Japan's tuna fisheries? A history of government-business relations in a food-producing sector. *Japan Forum* 20(2), pp. 139–70.

Barclay, Kate (forthcoming) A history of industrial tuna fishing in the Pacific Islands, in M. Tull and J. Christensen (eds), *History of marine animal populations: Asia Pacific region* [working title], MARE series (New York: Springer).

Bergin, A. and M. Harward (1996) *Japan's tuna fishing industry: A setting sun or a new dawn?* (New York: Nova Science).

Bestor, T. (1999) Constructing sushi: Culture, cuisine and commodification in a Japanese market, in S.O. Long (ed.), *Lives in motion: Composing circles of self and community in Japan*, pp. 151–90 (Ithaca: Cornell University Press).

Bestor, T. (2004) *Tsukiji: The fish market at the center of the world* (Berkeley: University of California Press).

Biggs, Stuart, Kanoko Matsuyama and Frederik Balfour (2011) Tsunami speeds "terminal decline" of Japan's fish industry. *Bloomberg Businessweek*, 24 April. Available at http://www.businessweek.com/news/2011-04-25/tsunami-speeds-terminal-decline-of-japan-s-fish-industry.html, accessed 20 May 2011.

Boyer, R. (2005) How and why capitalisms differ. *Economy and Society* 34(4), pp. 509–57.

Campbell, H.F. and R.B. Nicholl (1994) The economics of the Japanese tuna fleet, 1979–80 to 1988–89, in H.F. Campbell and A.D. Owen (eds), *The economics of Papua New Guinea's tuna fisheries*, pp. 39–52 (Canberra: Australian Council for International Agricultural Research).

Carpenter, Susan (2003) *Special corporations and the bureaucracy: Why Japan can't reform* (Basingstoke: Palgrave Macmillan).

Cerny, Philip G. (1997) Paradoxes of the competition state: The dynamics of political globalization. *Government and Opposition* 32(2), pp. 251–74.

Chen, Henry T. (2009) *Taiwanese distant-water fisheries in Southeast Asia 1936–1977*. Research in maritime history, No. 39 (Newfoundland, Canada: International Maritime Economic History Association, St John's).

Clift, Ben and Cornelia Woll (2012) Economic patriotism: Reinventing control over markets. *Journal of European Public Policy* 19(3), pp. 1–17.

Darier, E. (1999) Foucault and the environment, in E. Darier (ed.), *Discourses of the environment*, pp. 1–35 (Oxford: Blackwell).

DeWit, Andrew and Hiroko Yamazaki (2004) Koizumi's flawed political economy of decentralization. *ZMag*, 14 October. Available at http://www.zmag.org/content/showarticle.cfm?SectionID=17&ItemID=6418, accessed 22 November 2007.

Epstein, Charlotte (2003) WorldWideWhale: Globalisation and a dialogue of cultures? *Cambridge Review of International Affairs* 16(2), pp. 309–22.

Epstein, Charlotte (2008) *The power of words in international relations: Birth of an anti-whaling discourse* (Cambridge, MA: MIT Press).

Ferguson, James (1994) *The anti-politics machine: "Development", depoliticization and bureaucratic power in Lesotho* (Minneapolis: University of Minnesota Press).

Fisheries Agency (1995) The Kyoto Declaration and Plan of Action, International Conference on the Sustainable Contribution of Fisheries to Food Security, Kyoto, Japan, 4–9 December. Organised by the Government of Japan in collaboration with the Food and Agriculture Organization of the United Nations. Available at http://www.fao.org/DOCREP/006/AC442e/AC442e2.htm, accessed 3 June 2011.

Foucault, Michel (2007) *Security, territory, population. Lectures at the Collège de France, 1997–78*. G. Burchell (trans.), M. Senellart, F. Ewald and A. Fontana (eds), (Basingstoke, UK and New York: Palgrave Macmillan).

Francks, Penelope (2006) *Rural economic development in Japan: From the nineteenth century to the Pacific War* (London: Routledge).

Fujinami, N. (1987) Development of Japan's tuna fisheries, in D. Doulman (ed.), *Tuna issues and perspectives in the Pacific Islands region*, pp. 57–70 (Hawai'i: East-West Center).

Government of Japan (c.2003) Draft country note on fisheries management systems. Organization for Economic Cooperation and Development (OECD). Available at http://www.oecd.org/dataoecd/10/46/34429748.pdf, accessed 21 April 2011.

Government of Japan (2010) Fisheries of Japan – FY2009 (2009/2010) and fisheries policy outline for FY2010 (White Paper on Fisheries). Available at http://www.jfa.maff.go.jp/e/annual_report/2009/pdf/2009_jfa_all.pdf, accessed 21 April 2011.

Hall, P. and D. Soskice, eds. (2001) *Varieties of capitalism: The institutional basis of competitive advantage* (Oxford: Oxford University Press).

Hasegawa, Hiroyo (c.2010) Japanese food self-sufficiency and local initiatives to improve it. The International Society for Agricultural Meteorology. Available at http://www.agrometeorology.org/topics/environment-and-sustainability/japanese-food-self-sufficiency-and-local-initiatives-to-improve-it, accessed 15 June 2011.

Hayes, David (2010) Japan targets greater self sufficiency. *World Fishing and Aquaculture*, 3 November. Available at http://www.worldfishing.net/features101/new-horizons/japan-targets-greater-self-sufficiency, accessed 15 June 2011.

Hein, Laura E. (1994) In search of peace and democracy: Japanese economic debate in political context. *Journal of Asian Studies* 53(3), pp. 752–78.

Hirata, K. (2004) Beached whales: Examining Japan's rejection of an international norm. *Social Science Japan Journal* 7(2), pp. 177–97.

Hori, Takeaki (1996) *Tuna and the Japanese: In search of a sustainable ecosystem* (Tokyo: Japan External Trade Organization [JETRO]).

Ishii, Atsushi and Ayako Okubo (2007) An alternative explanation of Japan's whaling diplomacy in the post-moratorium era. *Journal of International Wildlife and Policy* 10, pp. 55–87.
Jessop, B. and S. Oosterlynck (2008) Cultural political economy: On making the cultural turn without falling into soft economic sociology. *Geoforum* 29, pp. 1155–69.
Johnson, Chalmers (1995) *Japan: Who governs? The rise of the developmental state* (New York: W.W. Norton and Company).
Kagoshima Prefecture Skipjack and Tuna Fisheries Cooperative Association (2000) *Kagoshima Ken Katsuo Maguro Gyogyō Kyōdō Kumiai Sōritsu Gojūnenshu Shi* (Kagoshima Prefecture Skipjack and Tuna Fisheries Cooperative Association fifty year history) (Tokyo: Suisan Shinshio Sha).
Koh, Sunhui and Kate Barclay (2007) Traveling through autonomy and subjugation: Jeju Island under Japan and Korea. *The Asia Pacific Journal: Japan Focus*. Available at http://www.japanfocus.org/-Kate-Barclay/2433, accessed 28 June 2011.
Luke, T. (1995) Sustainable development as a power/knowledge system: The problem of "governmentality", in F. Fisher and M. Black (eds), *Greening environmental policy: The politics of a sustainable future*, pp. 21–32 (London: Paul Chapman).
Luke, T. (1999) Environmentality as green governmentality, in E. Darier (ed.), *Discourses of the environment*, pp. 121–51 (Oxford: Blackwell).
Maclachlan, P.L. (2004) From subjects to citizens: Japan's evolving consumer identity. *Japanese Studies* 24 (1), pp. 115–34.
MAFF (2010) Shokuryo no kyoukyuu ni kansuru tokubetsu seron chousa no kekka ni tsuite (Results of Special Poll on Food Supplies). Office of the Minister of Agriculture, Forestry and Fisheries (MAFF), Japan, released 14 October. Available at www.maff.go.jp/j/press/kanbo/anpo/pdf/101014-01.pdf, accessed 25 July 2012.
MAFF (c.2005) What is "*Shokuiku*" (Food Education)? Ministry of Agriculture, Forestry and Fisheries (MAFF), Japan. Available at http://www.maff.go.jp/e/pdf/shokuiku.pdf, accessed 21 April 2011.
MAFF (2009) Result of the symposium 'Think about the World Food Security', 30 June. Update No. 720. Ministry of Agriculture, Forestry and Fisheries (MAFF), Japan. Available at http://www.maff.go.jp/e/maffud/2009/720.html, accessed 21 April 2011.
Makino, Mitsuaku and Hiroyuki Matsuda (2005) Co-management in Japanese coastal fisheries: Institutional features and transaction costs. *Marine Policy* 29, pp. 441–50.
Mashimo, Toshiki (2009) To what level could Japan's food self-sufficiency recover? Consumers Union of Japan website, 'Food Security' section. Available at http://www.nishoren.org/en/?p=287, accessed 21 April 2011.
MoFA (2009) Promoting responsible international investment in agriculture. Roundtable concurrent with the 64th United Nations General Assembly. Hosted by the Government of Japan in association with the World Bank, FAO, IFAD and UNCTAD, 29 September. Ministry of Foreign Affairs (MoFA), Japan. Available at http://www.mofa.go.jp/policy/economy/fishery/agriculture/investment.html, accessed 20 May 2011.
Mulgan, Aurelia George (2000) *The politics of agriculture in Japan* (London: Routledge).
Mulgan, Aurelia George (2002) *Japan's failed revolution: Koizumi and the politics of economic reform* (Canberra: Asia Pacific Press).
Mulgan, Aurelia George (2003) Agricultural policy and agricultural policymaking: Perpetuating the status quo, in J. Amyx and P. Drysdale (eds), *Japanese governance: Beyond Japan Inc*, pp. 170–93 (London: RoutledgeCurzon).
Mulgan, Aurelia George (2005) *Japan's interventionist state: The role of MAFF* (London: RoutledgeCurzon).
Ong, Aihwa (2006) *Neoliberalism as exception: Mutations in citizenship and sovereignty* (Durham, NC and London: Duke University Press).
OPRT (2011) Kesennuma, a major tuna fishing base, will be on a firm path to revival. Organization for the Promotion of Responsible Tuna Fisheries. *OPRT Newsletter International* 32, May. Available at http://oprt.or.jp/eng/oprt-news-letter/, accessed 22 December 2011.
Pinstrup-Andersen, Per (2009) Food security: Definition and measurement. *Food Security* 1, pp. 5–7.
Polanyi, Karl (1944) *The great transformation* (New York: Rinehart).
Renton, Alex (2005) One in ten fish is eaten in Japan. So why don't they know there's a shortage? *Observer Food Monthly*, 10 April. Available at http://www.guardian.co.uk, accessed 19 August 2010.
Ruggie, John Gerard (1982) International regimes. Transactions and change: Embedded liberalism in the postwar economic order. *International Organization* 36(2), pp. 379–415.

Shiraishi, Yuriko (1999) *All about fish – Gyo!* Brochure, available from Women's Forum for Fish (WWF) Secretariat, 3–12-15 Ginza, Chuo-ku, Tokyo 104–0061, Japan.
Smith, Roger (2004) Japanese whaling policy and food security. Chapter in unpublished D.Phil. thesis, St Anthony's College, Oxford University.
Smith, Roger (2008) Food security and international fisheries policy in Japan's postwar planning. *Social Sciences Japan Journal* 11(2), pp. 259–76.
Sumaila, U.R., A.J. Khan, A. Dyck, R. Watson, G. Munro, P. Tydemers and D. Pauly (2010) A bottom-up re-estimation of global fisheries subsidies. *Journal of Bioeconomics* 12, pp. 201–25.
Tarte, S. (1998) *Japan's aid diplomacy and the Pacific Islands* (Canberra: Asia Pacific Press).
WFS (2006) From the podium, Japan – Japon. World Food Summit (WFS), 13–17 November, Rome, Italy. Available at http://www.fao.org/docrep/003/x0736m/rep2/japan.htm, accessed 15 June 2011.
WTO (n.d.) 'Food security', in Glossary. World Trade Organization (WTO) website. Available at http://www.wto.org/thewto_e/glossary_e/food_security_e.htm, accessed 21 April 2011.
WTO (2005) Hong Kong WTO Ministerial Meeting 2005: Briefing notes, rules: AD, SCM including fisheries subsidies, negotiations to clarify and improve disciplines. Available at http://www.wto.org/english/thewto_e/minist_e/min05_e/brief_e/brief08_e.htm, accessed 21 April 2011.
WTO (2009a) Trade policy review. Report by Japan. World Trade Organization (WTO) Trade Policy Review Body WT/TPR/G/211, 14 January. Available at http://www.wto.org/english/tratop_e/tpr_e/tp311_e.htm, accessed 20 May 2011.
WTO (2009b) Statement by Mr Hirotaka Akamatsu, Minister of Agriculture, Forestry and Fisheries of Japan at the Seventh Session of the WTO Ministerial Conference, Geneva, 30 November–2 December. Available at http://www.wto.org/english/thewto_e/minist_e/min09_e/min09_statements_e.htm#il, accessed 20 April 2011.
WTO (c.2010) Introduction to fisheries subsidies in the WTO. Rules negotiations, World Trade Organization (WTO). Available at http://www.wto.org/english/tratop_e/rulesneg_e/fish_e/fish_intro_e.htm, accessed 20 May 2011.

Rare Earths: Future Elements of Conflict in Asia?

MING HWA TING

Independent Researcher

JOHN SEAMAN

Institut français des relations internationales

Abstract: *China's sudden suspension of rare earth exports to Japan in September 2010 represented the opening of a new front in the international competition for natural resources. As the demands of the global economy change, the international demand for rare earths, which are used in a diverse range of high-tech industries, has also increased. As China is currently the largest rare earths exporter, its actions in regulating rare earths exports will directly affect the interests of other states in the supply chain. This paper therefore examines how China assumed this dominant position in the supply chain. It also looks at how major rare earths consumers such as Japan and South Korea are reducing their reliance on and vulnerability to Chinese rare earths supplies through means such as seabed exploration in disputed territories, which might contribute to geopolitical tensions and instability in the East Asian region.*

Introduction

The sudden suspension of rare earth exports from China to Japan in September 2010 after a maritime dispute near the contested Senkaku/Diaoyu islands proved a rude awakening to an understudied problem. The issue of rare metal supplies had been largely overlooked for years, but it has now been thrust into the international spotlight. Apathy or a profound belief in the viability of free markets to resolve resource allocation issues allowed China to gain a near monopoly on the global production of rare earth elements, which are necessary components of many modern-day, high-tech products. China has been putting ever-stricter controls on the industry for years, notably through export permits and quotas, but there was still little political will abroad to face the issue until it was worthy of regular front-page news. The Chinese Government claims that export-limiting policies are driven by environmental concerns and the risk of depleting valuable natural resources. Meanwhile, China's official explanation was that any suspension of rare earth exports to Japan in September 2010 was due to the spontaneous and uncoordinated actions of various rare earth companies, traders or customs workers and that no formal directive was issued to this effect. However, in view of the tight control the Chinese Government appears to exercise over the sale and export of rare metals, this claim came across as rather weak and implausible (Ting, 2010a).

Until now, most attention on the competition for natural resources has been fixated on "visible" commodities such as oil and iron ore. There are clear signs, however, that with the shift towards a more high-tech and greening economy, there is an emerging front in the age-old competition for "invisible" natural resources that has thus far been overlooked. The names may vary, from "rare metals" to "technology metals", but the fact is that groups of once obscure elements have now become crucial components of modern industry and are increasingly becoming the new elements of geopolitical power. The rare earth group is only one piece of this larger category, but it is nevertheless an important case study that can give indications of the challenges that lie ahead. This paper examines China's unique position in the rare earth industry and how Japan and South Korea, two net importers of rare earth elements, have sought to reduce their vulnerability to China's new-found power. The analysis begins with a definition of some key concepts that are essential for understanding the issues. It then analyses China's place in the rare earth picture, how it obtained a near-monopoly, and the complexity of reasons for which it has sought to increase its control over the industry. Both Japanese and South Korean industries are directly threatened by China's actions. An examination of several components of their strategies to increase security of supply for rare earths reveals that China's policies have in many ways pressed its neighbours to forge cooperation among themselves and with other global players, and have also stirred other potentially thorny territorial issues in the form of seabed mineral deposits and the definition of Exclusive Economic Zones (EEZs).

Understanding rare earth elements: Key concepts

To properly analyse rare earth policies and their implications, it is first necessary to understand some key concepts. First, rare earth elements are a grouping of 17 distinct metals that are represented in the Periodic Table by the 15 elements of the lanthanide group (from lanthanum to lutetium [atomic numbers 57–71]) and include yttrium (39) and scandium (21). They are often sub-divided into light (including lanthanum, cerium, europium and neodymium) and heavy (dysprosium and yttrium) rare earths, in which case the "lights" are generally more abundant and the "heavies" less abundant. As with other rare or "technology" metals such as lithium, indium or gallium, these elements play an increasingly important role in the current high-tech economy. End products for rare earths range anywhere from consumer electronics to sophisticated weapons

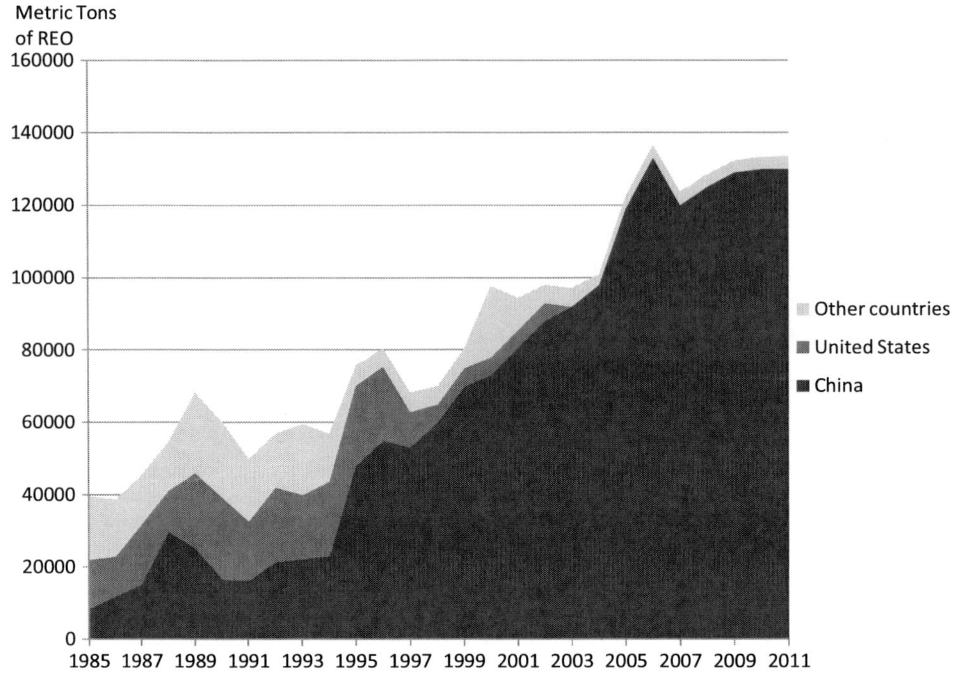

Figure 1. Estimated Global Mine Production of Rare Earths. Source: Metal Pages.

systems. Among the most cited industrial uses are the fabrication of clean energy products such as phosphors for energy-saving light bulbs or high-strength magnets used in the motors of hybrid and electric vehicles and the latest generation of wind turbines. As the drive for clean energy and improved energy efficiency gains momentum, various rare earth elements – in particular neodymium, dysprosium, europium and yttrium – will become increasingly important strategic resources as demand climbs to new heights (US Department of Energy, 2010). Furthermore, in the aftermath of the March 2011 nuclear disaster in Japan, international enthusiasm for nuclear energy has waned considerably, and one of the unintended consequences of this development is therefore the renewed push for renewable energy, which is likely to increase demand for rare earths (Ting, 2011).

Even though we come into contact with rare earths on a daily basis, they are used in relatively minute amounts compared to common materials such as iron and oil. The price of these metals relative to the cost of a final product is also quite small. As a result of the limited volume of rare earths in many products compared to other materials, their importance has largely gone under the radar (Ting, 2010b, p. 60). A good analogy would be to compare rare earths with spices. Spices are used sparingly in cooking, unlike common ingredients such as meat and potatoes. Hence, people frequently overlook the use of spices in cooking. This analogy also works in the sense that, more than 400 years ago, European powers were competing with each other to gain access to and control of spice-producing regions (Ting, 2010b, p. 58). Likewise, there are also strong indications that states are competing with each other for access to rare earth and rare earth-producing regions, which could become a potential flashpoint in the future.

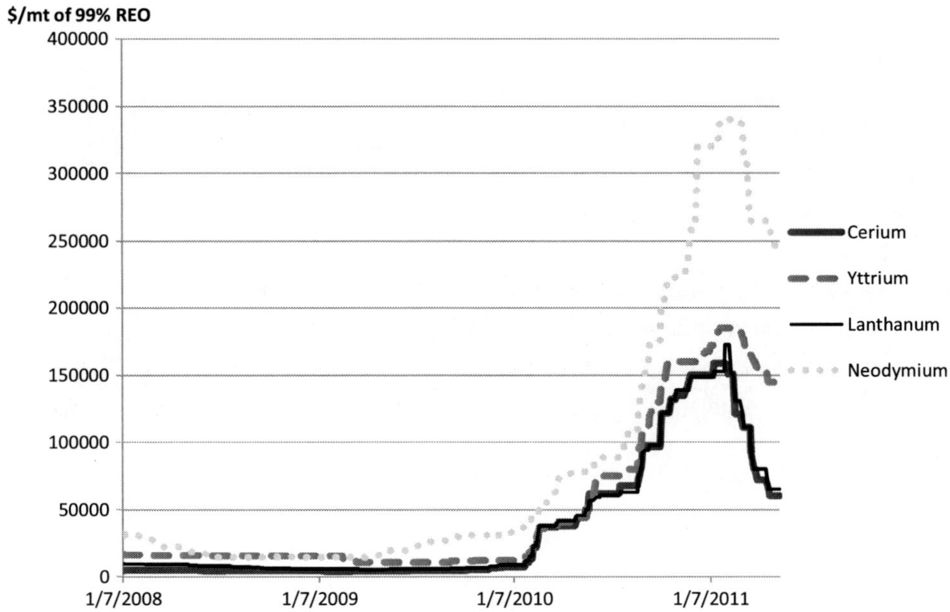

Figure 2. Export Prices of Select Chinese Rare Earths. Source: Metal Pages.

While exploitable rare earth deposits exist across the globe, bringing them into production and easing supply pressures is a complex endeavour. First, mining and refining these elements is a toxic process. Strong acids are used at various stages of separation and refinement, and radioactive materials such as uranium or thorium are often associated with minerals containing rare earths (Bradsher, 2011b). Severe health and environmental consequences for local populations and ecosystems could result from a failure to respect strict waste management procedures. This translates into production costs that are either internalised by the producer or externalised onto the local community.

Another salient feature of rare earths is the long lag times and high input costs from the discovery of an extractable deposit to the production of a useable oxide or concentrate. For instance, it takes seven to ten years to gain a mineral mining permit in the United States. This gestation period greatly hinders the reactivity of rare earth production to changes in market conditions. Getting rare earth production right therefore opens an investor up to a great deal of risk as prices can shift dramatically from the moment of initial investment to the moment a refined product goes up for sale. China's current dominant position in rare earth production, as well as the extent of government control, is therefore particularly troublesome to rare earth consumers abroad.

China and Rare Earths: A Monopoly Actor with Debatable Intentions

Leader of the pack: How China gained its monopoly

For much of the last decade, China has been the largest producer of rare earths, churning out more than 95 per cent of the world's rare earth oxides and concentrates. In 2010 its mines and smelters produced 130,000 tons of rare earth oxide (REO) out of a

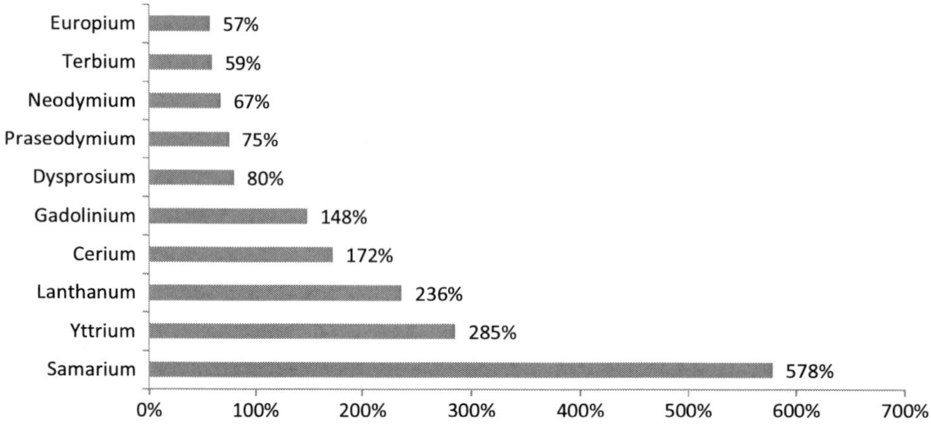

Figure 3. Price Markup of Selected Rare Earth Oxides. Source: Metal Pages.

global total of 137,000 tons (United States Geological Survey, 2011). This production has been largely limited to two primary regions, notably Baotou (Inner Mongolia) in the north, which produces high volumes of light rare earths but almost none of the heavies, and a scattering of operations in the southern provinces of Jiangxi, Fujian, Guangdong and Hunan, where heavy rare earth production is concentrated.

China is also the world's largest consumer of these commodities, accounting for more than 60 per cent of global demand. Despite its monopoly on production, China does not possess even a majority of the world's reserves and did not therefore obtain its position simply by natural disposition. According to the 2010 figures of the United States Geological Survey (USGS), China only possesses approximately 48 per cent of the world's known, economically exploitable reserves of rare earths – a proportion that is subject to some variation, as USGS figures from 2009 identified only 37 per cent of these reserves as being in China (United States Geological Survey, 2011, p. 129). Other major reserves exist in many places outside China, such as the United States and Australia (Hendrick, 2010, p. 129). Many industry experts even admit that comprehensive geological surveys that take rare earths into account have yet to be completed in much of the world and that data on viable mineral deposits is likely to evolve as surveying methods improve and markets and policies favour rare earth exploitation (Hatch, 2011).

The economic and strategic values of rare earths were recognised by the Chinese Government as early as 1990. They were officially classified as a protected and strategic commodity and fell under the purview of the State Development and Planning Commission and the Ministry of Commerce. When Deng Xiaoping visited the Chinese Rare Earths Research Institute in Baotou in 1992, he famously stated that, "There is oil in the Middle East; there are rare earths in China". Deng's prescient comparison of oil and rare earths shows that the former Chinese paramount leader was clearly aware that these commodities would play an increasingly prominent role in the future. By contrast, it would be decades before leaders in the rest of the world would come to the same realisation as Chinese policymakers. The West lacked a clear strategy and seemed content to let market forces do their work while global rare earth production and expertise rapidly shifted from the United States to China.

As production increased in China, mines overseas found it increasingly difficult to compete. In the United States in particular, stringent environmental regulations imposed on rare earth producers raised the cost of extracting and treating these materials, while a combination of cut-throat competition, relatively easy access and lax environmental regulation in China dropped the price of rare earths ever lower. As a result, from 2003 to December 2011, no rare earths were mined in the United States (United States Geological Survey, 2004, p. 132). The United States only re-started its rare earths mining operations when Molycorp started extracting rare earths ore from its Mountain Pass mine in December 2011.

By default, China became the dominant actor in this sector. Today, China is often accused of following a strategy of undercutting foreign producers with low prices in order to gain a monopoly position, "locking up" global rare earth production and thus controlling downstream industries (Hsiao, 2010). Despite the apparent support of the Government, the development of China's rare earth sector is anything but a textbook example of effective central planning. As recently as 2009, China accounted for over 100 individual rare earth producers, with many more small mines operating illegally (Chen, 2010). While a production quota cap of 89,200 tons of REO was set for 2010, real output exceeded this amount by more than 45 per cent, feeding a flourishing black market in rare earth trade (Tse, 2011, p. 4). Figures remain highly speculative, but Chinese media and industry analysts have estimated that anywhere from 30 to 50 per cent of demand outside China in 2011 was met by smuggling, which is often done by simply declaring products under a different name on customs forms (Wang, 2011). Chinese authorities have exercised little control over the production and export of rare earths and are now scrambling to get a hold on the industry by using draconian measures discussed below. This kneejerk reaction to gain control over the rare earth sector ultimately calls into question the idea that a monopoly on global rare earth production was really the intention of China's leadership. It was more the *lack* of Chinese control and the apathy of potential foreign producing countries that allowed China to become the world's default supplier. Furthermore, the measures that China is adopting to impose greater control on its industry serve to encourage overseas production and weaken China's dominant position in the long term.

China crashing the rare earths party

Even if achieving a global monopoly on rare earths was not the goal of Chinese leaders, its actions as the dominant actor have nevertheless been a cause of concern and debate. China has chosen to adopt a broad set of policy actions towards the regulation and control of this increasingly important sector that is causing potentially grave consequences for critical supplies to high-tech industries the world over.

Of chief concern is the establishment of ever-stricter export quotas. Since 2006 the Chinese Ministry of Commerce has set decreasing levels of export quotas, limiting and reducing the number of firms that are allowed to export rare earths in raw form. In 2006, 47 Chinese companies had permits to export rare earths; in 2010, only 22 of these companies were allowed to do so (Zhōng huá rén mín gòng hé guó shāng wù bù, 2005; 2009). In 2008, combined export quotas for both Chinese companies and foreign joint ventures totalled 56,000 tons, while in 2009 they totalled 50,142 tons, before being dropped again to a year-end total of 30,258 tons in 2010.[1] Within just two years, China had slashed its export quotas by roughly 26,000 tons, which translates into a cut of more than 45 per cent. This is more significant when compared to levels of rare earth demand outside China. While non-Chinese demand dipped in 2009 due to the global economic downturn, major rare earths consumers such as Japan and Korea were hit hard as demand in 2010 was estimated at 48,000 tons, or a difference of more than 35 per cent (Kingsnorth, 2010).

While export quotas are by far the most contentious policy measure, other trade distorting measures have also been implemented to discourage rare earths exports. Specifically, export taxes on rare earths were established in 2007 to complement export quotas. Originally set at between 15 and 25 per cent depending on the oxide or concentrate being exported, the rates on many more rare earth products were raised to 25 per cent in 2011 (Zhang and Ding, 2010a). Rebates on value added taxes of 17 per cent for exports on rare earth products have also been cancelled, adding a further disincentive to exports (Korinek and Jeonghoi, 2010).

Apart from controlling export volumes, China is also seeking to influence the pricing of rare earths. This would increase profitability for Chinese producers, and give Chinese authorities greater power to control global prices and even drive a wedge between domestic and international prices. Indeed, between July 2010 and July 2011, prices for various rare earth oxides jumped by anywhere from 300 to 500 per cent on the back of decreased supplies to foreign consumers. Since the summer of 2010, significant differences of anywhere between 50 and 500 per cent have also emerged between Chinese and international market prices for individual elements, giving clear incentive for foreign buyers to shift rare earth-dependent production to China. Export quotas and taxes have been the primary driver for this price gap, but industry representatives in Japan also claim that Chinese customs officials have engaged in price fixing at the border, effectively denying shipment of rare earth products if the indicated sale price does not meet a minimum level (Seaman, 2011a).

China has also sought to consolidate the industry and increase its global pricing power. In August 2010, it was reported that Inner Mongolia Baotou Rare Earth High Tech Co. and Jiangxi Copper Corp. (with rare earth mining rights in Sichuan Province) would be introducing a unified pricing mechanism (Levkowitz and Beauchamp, 2010, p. 5). After this announcement, China's State Council in September 2010 indicated that it would introduce new legislation to encourage mergers and acquisitions of existing rare earths companies, which in recent years have numbered as many as 123 mines and 73 processing firms (Zhang, 2011). The State Council announced in May 2011 that consolidation around the Inner Mongolia Baotou Rare Earth High Tech Co. in northern China was largely complete (Gov.cn, 2011a) and that consolidation in the south is planned for the next two years around three major producers, widely believed to be China Minmetals, Chinalco, and Ganzhou Rare Earth. A new industrial association with 155 members ranging from the mining sector to the smelting sector, which reports to the Ministry of Industry and Information Technology, was also created in April 2012 in order to foster greater oversight of the rare metals mining industry in China. To bring production under control and assist industry consolidation, the Ministry of Land Resources also began setting quotas on rare earth production in 2008, initially set at 119,500 tons, which was decreased annually to 89,200 tons in 2010 before being raised to 93,800 tons in 2011 (Hatch, 2010). The Ministry also placed a moratorium until 30 June 2012 on the issuing of new mining permits and at the time of writing it remains to be seen whether new permits will again be issued after this timeframe (Ministry of Commerce, 2011). Apart from regulating the rare earth industry through policy, China has also been stepping up efforts to stamp out smuggling (Gov.cn, 2011c). Rising global demand for rare earths, the imposition of export tariffs on different commodities and China's strict export quotas have boosted prices overseas, fuelling this illegal practice and undermining efforts by the Chinese authorities to gain greater control of the industry. After all, it is more profitable to procure these natural commodities domestically and then sell them abroad (Buckley, 2011).

Complex motivations…

The extent and nature of these measures are often interpreted as a concerted effort to choke off foreign industries, giving an unfair advantage to Chinese companies and attempting to coerce foreign manufacturers of high-tech equipment into relocating to China in order to access the necessary critical raw materials (Hurst, 2010, pp. 12–15). Although the point should not be dismissed, the reality is more complex. China's motivations are a combination of conserving its own natural resources, responding to severe environmental problems caused by careless mining, and promoting higher value-added growth and job creation in China by supporting the development of indigenous high-tech industries.

Environmental issues have also become a matter of growing concern in China. While it is true that the country's growth model has traditionally ignored environmental externalities, the Chinese population has become increasingly vocal in its discontent on these issues, albeit primarily on a local level. Despite the readiness of many overseas analysts to dismiss the environmental argument, China has stated on many occasions that any restrictions on the export of rare earths are motivated by the need for environmental protection and to ensure the long-term sustainability of this industry (2010). Indeed, local Chinese populations have borne the cost of serious environmental and health-related externalities while economies abroad have been rewarded with low prices (*People's Daily*, 2010). Many in China protest that rare earth mining has gone on for too long with no regulation, which has resulted in rare earths being sold for the price of Chinese cabbage (Lǐ Yī Péng, 2010).

Due to the lack of coordination between the mines and the Chinese Government, China's Ministry of Land Resources argues that over-exploitation of rare earths between 1996 and 2009 resulted in Chinese reserves declining by 37 per cent (Zhōng guó rén mín gòng hé guó guó tǔ zī yuán bù, 2010b). According to figures from the United States Geological Survey (1996; 2011), however, China's rare earths reserves increased from 43,000,000 tons to 55,000,000 tons between 1996 and 2010. Even though the supply of rare earths, like other natural resources, is finite, the increase in reserves is easily explained by the discovery of new deposits as well as improvements in technologies that allow the mining of previously inaccessible ores, thereby increasing both the supply and reserves. Yet the perception of rapidly diminishing reserves remains in China as it justifies its on-going control of this industry, particularly with the rarer, "heavy" rare earths, mined exclusively in China's southern provinces. Chinese scientists and policymakers claim that these mines cannot possibly keep up with global, or even Chinese, demand

and that China will become a net importer of some of these heavy rare earths by 2015, and will exhaust its reserves in the next 15 to 20 years (Seaman, 2011c).

In one effort to preserve the nation's natural resources and correct the deplorable environmental habits of the country's rare earth industries, firms now applying for permits to mine and export rare earth metals have to meet strict environmental standards, and must also have ISO 9000 certification (2010). The Ministry of Land Resources has made it clear that it favours erecting increased barriers to entry for future rare earth operations in order to increase regulation of this industry by reducing the number of companies involved (Zhōng guó rén mín gòng hé guó guó tǔ zī yuán bù, 2010a). Consolidating industries in China is nothing new and has been used on many occasions to rein in inefficient, polluting industries – small, ineffective coal-fired power plants being one prominent example. Indeed, China redoubled efforts in 2011. The Ministry of Environment announced in April a series of tougher environmental standards to take effect from October 2011 (Wū raõn fáng zhì sī, 2011).[2]

An inspection plan to enforce these standards was also announced, with local environmental protection bureaus expected to submit preliminary reports in September that would be followed by inspections from Beijing officials before December 2011 (Gov.cn, 2011b; Hook, 2011). The results of these inspections ultimately translated into export quota allocations for 2012. Producers who met Government standards were allotted "confirmed" quotas while those who did not were given "provisional" quotas that could be reassigned if a given producer continued to fall short (Hatch, 2011b). Ultimately, there are considerable costs associated with adhering to new environmental standards. Measures to support prices in China are therefore seen as crucial to this initiative and all efforts used to bring greater control to the industry, be they quotas, taxes or other restrictions, are tools for achieving a cleaner, more sustainable industry (Lóng Huí Rén, 2010).

Yet there is another side to the environmental coin. Resorting to the need for environmental protection shields China from international demands for increased output and criticisms of restricting exports. China's export quotas fly in the face of its obligations under the concession agreement of the World Trade Organization (WTO), which it joined in 2001 and which expressly forbids the implementation of export quotas. The only exception, listed in Article 20 of the WTO charter, is quotas destined for the protection of the environment and scarce natural resources. It would seem convenient for China to introduce strict environmental standards on the rare earths industry just as the global demand for these materials is increasing. Having driven foreign producers out of the market with lower prices, it is conceivable that the timing of switching to a more sustainable model of extraction just when demand is increasing has been motivated by economic or geopolitical considerations rather than solely environmental concerns. The credibility of this scenario is supported by recent talks between China and Taiwan that would see the latter obtaining preferential access to rare earths exports from China, a move that appears to contravene WTO regulations (Mozur and Liu, 2011). Furthermore, China's Baotou Rare Earths High-Tech Co., the top rare earths producer, announced in October 2011 that it was suspending its smelting and separation processes for a month in order to "stimulate the market" (Lian and Stanway, 2011); and no environmental concerns were expressed in its filing to the Shanghai Stock Exchange. This temporary suspension came in the wake of a 19 May statement released by the Central Government that made explicit the need for it to exercise leadership in the rare earth industry. Apart from raising the barriers of entry for rare earths exporting companies, the statement also noted the need to have in place comprehensive taxation and quota policies in order to achieve uniform pricing (Zhōng huá rén mín gòng hé guó zhōng yāng rén mín zhèng fǔ, 2011). Consequently, the United States, the European Union and Japan lodged a formal complaint against China's restrictions on rare earths exports in the WTO in March 2012. Significantly, this effort represented the first time Japan has formally complained against China. Not only does this show the degree of importance the Japanese Government attaches to having equitable access to such commoditiess; it also shows that Japan is not "turning a blind eye to China's further disobedience of global rules" (*Yomiuri Shimbun*, 2012).

Indeed, a major pillar in the Communist Party's legitimacy is its ability to guarantee economic growth and prosperity in a shifting economic climate. At home, upward pressure is being put on wages while droves of young, qualified workers are entering the Chinese job market each year. At the same time, the effects of the international financial crisis have also called into question the viability of overseas markets and China's dependence on exports for growth, necessitating an increase in internal consumption, which ultimately demands higher incomes that come from higher value-added economic activities. While trade in rare earths was worth no more than 2 billion USD in 2010, the downstream industries that depend on them are worth far, far more (Kingsnorth, 2010). Zhao Shuanglian, Vice Chairman of the Inner Mongolia Autonomous Region, said that the objective was "to use moderation in the control of the production of rare earth resources and reduce exports to an acceptable level [so as] to attract more Chinese and foreign investors into the region" (*China Daily,* 2009). In effect, such a policy also allows China to draw in foreign investment, downstream industries, and technology as foreign firms would have to conduct their research and development in China if they wanted guaranteed supplies of rare earths. Squeezed by high prices and a lack of raw materials as commercial stockpiles began to reach their limit, many Japanese companies began to build factories in China. In the magnet industry alone, alloy producers Showa Denko and Santoku have established operations in southern China to access heavy rare earths while Hitachi Metals, which holds a number of key patents on magnet production until 2014, is thought to be looking into investments in China as well. Similar stories abound in industries that depend on rare earths for phosphors, batteries, catalysts and a host of other components of high-tech products (Inoue and Gordon, 2011).

Nevertheless, it should be underlined that China's policy actions have in fact resulted in a favourable pricing environment, not to mention plenty of political backing, to develop rare earth production overseas. Rare earth prices outside China have soared since the summer of 2010, ultimately compensating for the high costs of opening new mines overseas (Bradsher, 2011a).

Some investors and industrial groups argue that China's power in the market discourages investments, despite the higher price. Indeed, China has the power to remove its export quotas, sending prices falling and undercutting the profits of overseas producers once again. But such a move is highly unlikely given China's growing need for rare earths at home. According to Chen Zhanheng of the Chinese Society of Rare Earths, Chinese demand for rare earths is climbing rapidly and China is likely to become a net importer of certain rare earths by 2015 (Chambers, 2011). Industry and policy experts in China note that even if Chinese policymakers wanted to undercut overseas producers, Chinese demand for rare earths is so great that it must maintain its export quotas in order to guarantee supplies for its own industries (Seaman, 2011d).

Japan, South Korea and the Search for Alternatives

Regardless of China's intentions, its actions have effectively reduced the supply of rare earths to industrial users overseas and forced both the private sector and public policymakers abroad to press for solutions to reduce their dependence on Chinese rare earths. Japan has been developing its policy on this issue for years, while South Korea has begun to reflect on its longer term industrial strategies and potential vulnerabilities. An analysis of Japanese and South Korean actions for dealing with the rare earth issue reveals some interesting trends. China's policies have pressed both countries to deepen their ties with other rare earth producers in the region and beyond, with developed economies such as the United States and the European Union, and with each other. As this cooperation has occurred as a direct response to what are seen as aggressive Chinese policies, Chinese actions have essentially served to isolate China in this specific case. Another unsettling consequence of the rare earth issue is the potential for opening new chapters in territorial disputes between these East Asian neighbours regarding Exclusive Economic Zones (EEZs) and seabed deposits of various rare metals.

Mapping vulnerabilities and threat perceptions

As an economy with established high-tech industries that constitute an essential pillar of its economic activity, Japan is the most dependent country on rare earths outside China. In 2010, it imported an estimated 30,000 tons of rare earths, accounting for more than half of global consumption outside China. Chinese policies thus affect Japan disproportionately. Its vulnerability to supply disruptions has been a source of concern for many years, long before the territorial dispute in the East China Sea brought the issue to the world's attention. In order to reduce the possible impact of another episode of export restrictions, the Japanese Diet voted a supplementary budget in the fall of 2010 that provided the Ministry of Economy, Trade and Industry (METI) with an additional 100 billion yen (1.2 billion USD) in 2011 to develop a range of alternatives to Chinese rare earths (Ministry of Economy, Trade and Industry, n.d.).

South Korea, with its budding automotive and consumer electronics industries, is also finding its demand for rare earth elements growing. Imports of these elements totalled only 2,655 tons in 2009, with 65 per cent of these imports coming from China, while the rest were delivered via Japan (Lim, 2010). This figure is likely to rise as rare earth-dependent industries on the peninsula expand. South Korean companies such as Samsung and LG will need continual and fair access to rare metals as they seek to establish themselves as global technological leaders. Likewise, automobile companies such as Hyundai and Kia require access to rare metals and rare earths in particular to develop hybrid and electrical vehicles. Consequently, South Korea is also stepping up its research in rare metals. For instance, speaking to the *Korea Times*, Na Kyoung-hoa, president of the Korean Institute of Industrial Technology (KITECH), said: "It is critical that we secure rare metals as they are essential for the nation's high-tech and green industries" (Han, 2010). Korea recognises the inherent risks associated with being over-reliant on Chinese sources and there has been a push to diversify its supplies of rare metals and rare earths more specifically (Kim, 2010). In November 2009, the South Korean Government announced that it is setting aside 300 billion won (250 million USD) to develop 40 core rare metals industries and improve its self-sufficiency ratio on rare metal refining from 18 per cent to 80 per cent by 2018 (Kim, 2009).

Scouring the globe for rare earths – enhancing bilateral cooperation in the face of China

Diversifying supplies of rare earths by signing partnerships with countries or firms overseas has been an essential part of private- and public-sector policy in both countries to ensure security of supply for high-tech industries. This push to diversify supplies away from China has served to reinforce both bilateral and multilateral cooperation.

Expectedly, Japan has been actively seeking out alternative supplies of rare earths. Companies such as Sojitz, Toyota, Mitsui, Hitachi and Sumitomo have been working to secure rights to explore, exploit or import rare earths from Asia, Africa and the Americas. Corporations such as Toyota Tsusho and Sojitz have also signed production agreements with Vietnam to develop a mine in Lai Chau that could produce up to 3,000 tons of rare earths annually (Kubota, 2010). Meanwhile, during the dispute in the East China Sea, Indian Prime Minister Singh and Japanese Prime Minister Kan reached an agreement to explore rare earth cooperation. Six weeks later, Toyota Tsusho would agree to construct a rare earths processing plant in India with Indian Rare Earths. In February 2011, while Japan and India were negotiating a historic free trade agreement, ministers from both sides agreed to further cooperation in exploring Indian rare earth deposits (Monahan, 2011). Indeed, rare earth cooperation has been a non-negligible factor in Japan's latest efforts to develop enhanced trade agreements with its Asian partners (Mulgan, 2011).

Australia has also figured high on Japan's rare earth priority list. In November 2010, shortly after the Sino-Japanese territorial dispute, Australian mining company Lynas agreed to sell Sojitz 8,000 to 9,000 tons of rare earths a year over a 10-year period in exchange for a 250 million USD loan (Fickling, 2010). Lynas owns what is billed as the largest rare earth mine outside China. Production was planned to start in the fall of 2011 but as of early 2012 environmental permits had yet to be issued for the project's refining operations in Malaysia, as local concerns over lax environmental safeguards have turned political. Lynas was slated to appeal to the Minister of Innovation, Science and Technology in late April 2012; operations can only move forward if the appeal is successful. Nevertheless, a key feature of this, and most Japanese rare earth deals, is the financial support granted by the Japanese state via the Japan Oil, Gas and Metals National Corporation (JOGMEC). JOGMEC is using special funds allocated to rare earth projects by the October 2010 stimulus package and, following a recent change in Japanese law, is allowed to participate directly in a mining venture in addition to providing financial backing for Japanese companies and technical expertise in the field (Sojitz Corporation, 2010).

Given their shared vulnerability to fluctuations in the Chinese supply chain, Japan has also been collaborating with South Korea to explore the viability of alternative sources. According to a December 2010 press release from the Korean Ministry of Knowledge Economy, Vice Minister for Trade and Energy Park met with senior Japanese officials to discuss future collaboration regarding the joint development of mines, research into substitutes and recycling of rare earths (Ministry of Knowledge Economy, 2010). In a sign of close cooperation between them, Japanese and Korean firms bought a 15 per cent stake in Companhia Brasileira de Metalurgie e Mineracao (CBMM) (BBC, 2011). They have also teamed up with firms from the United States to invest in rare earth mining development in Mongolia, where the local partner Rare Earth Exporters of Mongolia – a joint venture partner of Green Technology Solutions Inc. (GTSO), an Australian company – portrays itself on its website as a "checkmate on China" (Rare Earth Exporters of Mongolia). GTSO shipped its first Mongolian rare earth core samples to South Korea in September 2011 to be analysed by the state-owned Korea Resources Corp. (KORES) (2011).

As a further indication of South Korea's firm intent to wean itself off Chinese supplies, it has also entered into another agreement, this time with Cameroon in April 2011, to undertake joint development and exploration of the latter's rare earth resources. This deal follows an announcement in January 2011 by the South Korean Government in relation to securing natural resources vital to its economic development, and to triple its funding to 7 billion USD to secure natural resources such as oil, gas and rare earths from overseas sources. As a testament to South Korea's commitment in ensuring that it intends to be competitive on the rare earth front, the Government also announced in October 2010 that it is planning to stockpile 76,000 tons of rare metals and rare earth elements by 2016 (Yonhap News, 2010). In the second half of 2011, South Korea's first storage facility, costing 17 billion won, with enough capacity to store 35-day consumption of these strategic resources, will come into operation. In the short term, the plan is to increase current stockpiles which would last 8.1 days to 13.5 days (Cho, 2010).

Seabed deposits – a possible source of contention

Apart from securing foreign supplies through more traditional methods discussed above, Japan is also looking into extracting its own limited supplies of such raw materials using innovative methods. In a bid to reduce its dependency on Chinese exports, Yukio Hatoyama's Cabinet was seriously considering a proposal to undertake concerted action to extract rare earths from the seabed within Japan's Exclusive Economic Zone (EEZ), a move that was continued by the Naoto Kan and Yoshihiko Noda administrations. Japanese efforts to extract commodities from seabeds are not surprising since it was looking to the seas for other natural resources even before the recent attention on rare metals. As early as 1982, the Natural Resources and Energy Agency of the Ministry of International Trade and Industry had made clear its intentions to extract rare metals from the bottom of its seas (1982). In June of the same year, Japan announced plans to implement a reserves system to stockpile rare metals starting in April 1983 (The Japanese government is planning to stockpile a range of rare metals, 1982). As recently as March 2007, the Japan Agency for Marine-Earth Science and Technology announced that underwater probes had discovered deposits of various metals such as manganese, copper, zinc and lead, as well as methane hydrates. A team of researchers led by Yasuhiro Kato of the University of Tokyo also reported significant findings of rare earth elements in seabed mud thousands of metres below the surface of international waters in the Pacific Ocean (Kato et al., 2011). While this latest discovery succeeded in creating a media frenzy, many industry representatives are nevertheless pessimistic about the prospects of commercialising such an endeavour, as uncertainties about the environmental impact are yet to be resolved, capital costs remain exorbitant and any real operational project is thought to be 10 years away at the very least. Seabed ferro-manganese nodules, crusts and mud are shown to contain high concentrations of rare earths, but land-based solutions are seen as more secure investment opportunities (Seaman, 2011c).

Nevertheless, the prospect of a rich, indigenous source of rare earths and other raw materials could be a strategic boon for the traditionally resource-poor country and is driving seabed exploration. To this end, Japan has also signalled its intention to devote significant resources towards developing infrastructure at Minamitorishima and Okinotorishima. On 14 January 2010, the Japanese Government authorised the appropriation of funds to build sea walls there, which incidentally are Japan's southernmost and easternmost maritime features. According to Japan, the ongoing urban renewal efforts are to prevent coastal erosion; cementing its EEZ claims is not the professed objective, though they are directly strengthened by the exercise of sovereignty. In preparation for the start of scouring the seabed for rare metals, plans are also under way to construct ports at these two outlying islands (Ting, 2010b, p. 63). China has already expressed its opposition since it does not recognise the Japanese EEZ claims. To China, Okinotorishima is merely a rock; to Japan, it is an island. The distinction marks the difference between sovereign right over the territorial waters within 12 nautical miles of the atoll, and an EEZ with a radius of 200 nautical miles. Further complicating matters, Okinotorishima's relative proximity to Taiwan means that Japan's ongoing urban renewal efforts to strengthen and increase its presence there are unlikely to please China, especially if rare metals are then extracted from the disputed seabed. Furthermore, in the event of a conflict between China and Taiwan, China requires open sea lanes to conduct its naval activities between the East China Sea and the Pacific Ocean. Hence, any unilateral efforts by Japan to bolster its claims over Okinotorishima are unlikely to be welcomed by China. Moreover, China has already made it clear that it opposes Japanese efforts at scouring disputed seabeds for minerals (Yang, 2010). Given the contested nature of the area surrounding the seabeds, any unilateral actions, especially large-scale drilling by Japan without prior consultation with China, might be misinterpreted by the latter to be belligerent actions, which might lead to increased tensions between them. Likewise, China's plan to undertake underwater geological surveys in the East China Sea is likely to lead to tension with Japan as both actors have overlapping EEZ claims in the region (2011). As a result, the competition for such rare earths may prove to be a catalytic element of conflict between these two East Asian neighbours in the future.

Searching for innovative solutions

Apart from extracting metals from the environment, Japan has begun concerted efforts to reduce rare earth demand by developing substitute technologies and improving the efficiency of resource use. METI's supplemental budget for 2011 included 12 billion yen

(145 million USD) for such research and development, while the Ministry of Education has been busy developing a 10-year program to find technological solutions to the rare earth supply problem, notably in the fields of magnets, catalysts and batteries (Seaman, 2011e). Companies such as Toyota, for example, have been researching new hybrid and electric motors that do not require rare earth magnets (Ohnsman, 2011), while magnet producers have been researching different ways to reduce the amount of rare earths required and to recycle industrial scrap to avoid wasting resources (Seaman, 2011b).

On the supply side, pilot recycling projects for high-tech products such as laptops, personal computers, fluorescent lights and monitor screens are currently under way in Japan to extract rare earths and other metals present in the various components (Nagoya International Center, 2010). Apart from the city of Nagoya, Kosaka city has also jumped on the recycling bandwagon (Tabuchi, 2010). Previously, collection efforts in Japan were directed mainly at ensuring the proper disposal of end-of-life consumer appliances as opposed to the extraction of recyclable rare metals. Such electronic waste has traditionally been collected and shipped to developing countries for disposal in landfills or to be stripped for parts (Barnes, 2011). There are promising signs that the domestic electronics recycling policy is paying dividends in Japan. In December 2010, Hitachi Ltd. announced that it had developed new technology and equipment that allows for the extracting and recycling of rare earth magnets from consumer appliances such as hard disk drives, air conditioners and other types of compressors. This automated process is more efficient than if it were to be done manually. Hitachi is expected to commence full-scale recycling in 2013 and meet 10 per cent of its needs at that time (Hitachi, 2010).

Conclusion

Rare earth elements are ultimately one example of an understudied set of natural resource issues that both industrialised and industrialising countries will increasingly be called upon to face. Advances in technology breed new sources of economic growth and prosperity, but also generate new dependencies on a broad range of either scarce or concentrated raw materials, be they called "technology", "critical" or simply "rare" metals. Questions over access to these resources could easily become a source of tension, or in an extreme case, conflict. If production is unregulated, this new resource binge may spell disaster for local environments and human health in mining regions. Yet, on the other hand, those countries that find themselves in a privileged supplier position may be tempted to use their resource advantage to benefit their own industries at the expense of foreign demand. In the end, policies that act on both of these dimensions serve to cloud the real intentions of the actors involved and lead to mutual misunderstandings and a polarisation of the issues. The above analysis of Chinese, Japanese and, to an increasing extent, South Korean policies and actions over rare earths helps to illustrate how emerging rare metal dependencies can play a role in increasing regional instability in Asia.

As the world's monopoly producer of rare earths, China has been implementing a number of measures to consolidate the industry and exert greater control over these increasingly important commodities. These policies have been driven in part by an effort to benefit fully from the strategically advantageous position China finds itself in

– using its control over resources to help drive technological advancement and create within China higher value-added links in high-tech production chains.

In recent years, environmental concerns have gained considerable importance in Chinese political calculus as its "pollute first–ask later" model seems to be reaching its limit. The Chinese Government's stated goal of cleaning up a highly toxic industry and preserving national treasures cannot therefore be fully dismissed as a core motivation for reining in wasteful mining and processing operations, even if the measures have been considered draconian and have led to far-reaching international consequences. Indeed, industrialised economies have particularly felt the impact of Chinese export restrictions, not only through rising prices but also through a drastic reduction in simple access to rare earths. Whether intentional or not, the near stoppage of rare earth exports to Japan in the final quarter of 2010 is a clear example.

Japan and, more recently, South Korea have been vigorously searching for ways to reduce their dependence on Chinese rare earths, hoping to secure supplies through partnerships with other producers overseas, by finding accessible deposits domestically, or by innovation in recycling and industrial efficiency techniques. Ultimately, one consequence has been to open new chapters in territorial disputes over maritime claims and EEZs as potentially exploitable deposits are explored along the seabed. Another significant consequence is that these efforts have brought new opportunities for regional cooperation where resistance to China's monopoly position is a rallying call. Regardless of the real intentions behind China's rare earth policies, the ultimate effect has been to draw suspicion that has left the country further isolated and the region incrementally more unstable.

Notes

1. All figures are based on figures released by the People's Republic of China Ministry of Commerce and gathered from Metal Pages. The URLs are http://english.mofcom.gov.cn/ and http://www.metal-pages.com/ respectively. These websites were accessed in November 2011.
2. For more information, see Xī tǔ gōng yè wū raõn wù pái fàng biāo zhǔn (GB26451 fàng biāo), http://kjs.mep.gov.cn/pv_obj_cache/pv_obj_id_41D98CE00CFE396700315824DA96D4F024250600/filename/W020110210366768105784.pdf, accessed November 2011.

References

Barnes, Abi (2011) New war on e-waste. *China Dialogue*, 10 August.

BBC (2011) Japanese and Korean firms look to Brazil for rare metal, 3 March.

Bradsher, Keith (2011a) Supplies squeezed, rare earth prices soar. *New York Times*, 3 May.

Bradsher, Keith (2011b) Taking a risk for rare earths. *New York Times*, 8 March.

Buckley, Chris (2011) China vows to crack down on rare earth smuggling. *Reuters*, 19 May.

Business Wire (2011) GTSO ships Mongolian rare earth core samples to South Korea. *Business Wire*, 12 September.

Chambers, Matt (2011) China, biggest produce [sic] of rare earths, expects to import more. *The Australian*, 5 February.

Chen, Zhanheng (2010) *Outline on the development and policies of China rare earth industry*. The Chinese Society of Rare Earths (CSRE). Available at http://www.cs-re.org.cn/en/modules.php?name=News&file=article&sid=35, accessed 12 May 2011.

China Daily (2009) Exports control on rare earths aims at growth. *China Daily*, 2 September.

China Daily (2010) Ministry begins accepting applications for rare earth export quotas, 12 November.

China steps up regulation of rare earth industry. Available at http://english.gov.cn/2011-09/27/content_1957496.htm, accessed 27 September 2011.

Cho, Mee-young (2010) Update 1–S. Korea to lift 2011 energy, resource development. *Reuters*, 15 December.

Fickling, David (2010) Lynas, Solitz to market, distribute up to 90,000 tonnes of rare earths. *The Australian*, 24 November.

Gov.cn (2011a) *China issues guideline to promote healthy development of rare earth industry*. Gov.cn. Available at http://english.gov.cn/2011-05/19/content_1867140.htm, accessed 19 May 2011.

Gov.cn (2011b) *China launches campaign to regulate rare earth production*. Available at http://english.gov.cn/2011-08/09/content_1922201.htm, accessed 9 August 2011.

Gov.cn (2011c) China steps up regulation of rare earth industry. Available at http://english.gov.cn/2011-09/27/content_1957496.htm, accessed 27 September 2011.

Han, Jane (2010) Rare metal research center opens. *Korea Times*, 28 January.

Hatch, Gareth (2010) *More detail on the 2011 Chinese rare earth production quota*. Technology Metals Research. Available at http://www.techmetalsresearch.com/2011/04/more-detail-on-the-2011-chinese-rare-earth-production-quotas, accessed 17 April 2011.

Hatch, Gareth (2011a) *USGS publishes 2011 estimate of global rare-earth reserves*. Technology Metals Research. Available at http://www.techmetalsresearch.com/2011/02/usgs-publishes-2011-estimate-of-global-rare-earth-reserves/, accessed 20 May 2011.

Hatch, Gareth (2011b) *The first round of Chinese export quota allocations for 2012*. Technology Metals Research, 28 December. Available at http://www.techmetalsresearch.com/2011/12/the-first-round-of-chinese-rare-earth-export-quota-allocations-for-2012/, accessed 10 April 2012.

Hendrick, James B. (2010) Rare earths, in Mineral Commodities (ed.), *Summaries 2009* (Reston: United States Geological Survey).

Hitachi Ltd. (2010) *Hitachi develops recycling technologies for rare earth metals*. Hitachi Ltd. Available at http://www.hitachi.com/New/cnews/101206.html, accessed 6 December 2010.

Hook, Leslie (2011) Beijing crackdown hits rare earth mining. *Financial Times*, 23 August.

Hsiao, Russell (2010) Strategic implications of China's consolidation of rare earth industries. *China Brief* 20.

Hurst, Cindy (2010) *China's rare earth elements industry: What can the West learn?* Institute for the Analysis of Global Security.

Inoue, Yuko and Julie Gordon (2011) Japanese rare earth consumers set up shop in China. *Reuters*, 12 August.

Kato, Yasuhiro, Koichiro Fujinaga, Kentaro Nakamura, Yutaro Takaya, Kenichi Kitamura, Junichiro Ohta, Ryuichi Toda, Takuya Nakashima and Hikaru Iwamori (2011) Deep-sea mud in the Pacific Ocean as a potential resource for rare-earth elements. *Nature Geoscience* 4(8), pp. 535–39.

Kim, Hyun-choel (2009) Korea moves to secure rare metals. *Korea Times*, 27 November.

Kim, Young-gyo (2010) S. Korea urged to diversify source of rare earth metals. *Yonhap News*, 30 December.

Kingsnorth, Dudley (2010) *Meeting the challenges of rare earth supply in the next decade* (The Hague: Center for Strategic Studies).

Korinek, Jane and Jeonghoi Kim (2010) Export restrictions on strategic raw materials and their impact on trade and global supply, in *The economic impact of export restrictions* (Paris: OECD).

Kubota, Yoko (2010) Vietnam and Japan to mine rare earths together. *Reuters*, 31 October.

Levkowitz, Lee and Nathan Beauchamp (2010) China's rare earths industry and its role in the international market, in *US–China Economic Security Review Commission Staff Backgrounder* (US–China Economic and Security Review Commission).

Lǐ Yī Péng, Rèn Mī Nuó and Shí Chàng (2010) Zhōng guó xī tǔ yuán hé mài le; liaõo;liào "bái cài jià". Rén mín rì bào haõi wài baõn.

Lian, Ruby and David Stanway (2011) China's Baotou rare-earth suspend facilities for one month. *Reuters*, 17 October.

Lim, Bomi (2010) South Korea to cut China rare-earth dependency as imports drop. *Reuters*, 22 October.

Lóng Huí Rén (2010) Quán guó rén dà dài biaõo liú jī fú: jiā qiáng duì xī tǔ zī yuán de baõo hù. Available at http://www.mlr.gov.cn/kczygl/kcbh/201103/t20110310_822950.htm, accessed 22 May 2011.

Ministry of Commerce, China (2011) *China caps rare metals output, raises mandatory output*. Available at http://english.mofcom.gov.cn/aarticle/newsrelease/counselorsoffice/westernasiaandafricareport/201104/20110407527218.html, accessed 20 May 2011.

Ministry of Economics, Trade and Industry, Japan (n.d.) 鉱物資源　石炭開発を巡る最近の動向 (Tokyo: METI).

Ministry of Knowledge Economy, Korea (2010) *Korea, Japan take joint action to secure rare earths*. Available at http://www.mke.go.kr/language/eng/news/news_view.jsp?seq=972&srchType=1&srchWord=&tableNm=E_01_01&page No=1, accessed 15 March 2011.

Ministry of Science and Technology, Japan. *S&T program*. Available at http://www.most.gov.cn/eng/programmes1/index.htm, accessed 15 March 2011.

Monahan, Andrew (2011) India and Japan look to expand rare earth alliance. *Wall Street Journal*, 16 February.

Mozur, Paul, and Fanny Liu (2011) Taiwan, China discuss possible rare-earths deal. *Wall Street Journal*, 19 May.

Mulgan, Aurelia George (2011) No longer a "reactive state": Japan's pro-free trade posture, in *Asie Visions* (IFRI).

Nagoya International Center (2010) *Recycling for rare metals*. Available at http://www.nic-nagoya.or.jp/en/e/2010/09/09/recycling-for-rare-metals-2/, accessed 10 December 2010.

Ohnsman, Alan (2011) Toyota readying motors that don't use rare earths. *Bloomberg*, 14 January.

People's Daily (2009) China's Inner Mongolia regulates rare earth export to attract investment: Official. *People's Daily*, 2 September.

People's Daily (2010) China: Restrictions measures target rare earth protection. *People's Daily*, 28 October.

People's Daily (2011) Double standards of Western companies. *People's Daily*, 11 January.

Rare Earth Exporters of Mongolia (n.d.) Rare Earth Exporters of Mongolia. Available at http://www.rareearthexporters.com, accessed 15 May 2011.

Rathi, Akshat (2011) Smuggling key factor in China's rare earth actions. *Chemistry World*, 29 October.

Seaman, John (2011a) Interview with Japanese industry representatives, Tokyo.

Seaman, John (2011b) Interview with Toru H. Okabe, Okabe Laboratory, University of Tokyo, Tokyo.

Seaman, John (2011c) Interviews with anonymous industry representatives, Tokyo.

Seaman, John (2011d) Interviews with industry and policy experts, Beijing.

Seaman, John (2011e) Interviews with Japanese Government officials.

SIJI Press English News Service (1982) Japan will go all out in exploring seabed mineral resources on its own without joining a four-nation Western agreement on the development of manganese nodules in the Pacific. *SIJI Press English News Service*, 18 February.

Sojitz Corporation (2010) Agreement to form strategic alliance with Lynas, an Australian rare earths company, concerning rare earths supply for the Japanese market and supporting the expansion of Lynas Rare Earths Project. Available at http://www.theaustralian.com.au/business/mining-energy/lynas-sojitz-to-market-distribute-up-to-90000-tonnes-rare-earths/story-e6frg9df-1225960333556, accessed 30 December 2010.

Tabuchi, Hiroko (2010) Japan recycles minerals from used electronics. *New York Times*, 4 October.

Ting, Ming Hwa (2010a) China and the supply chain of rare metals: Table of [dis]contents. *East Asia Forum*. Available at http://www.eastasiaforum.org/2010/11/11/china-and-the-supply-chain-of-rare-metals-table-of-discontents/, accessed 14 May 2011.

Ting, Ming Hwa (2010b) New spice wars: China, Japan, the US compete for rare metal. *Global Asia* 5(2), pp. 58–59.

Ting, Ming Hwa (2011) *Rare metals after the Japanese nuclear crisis*. Available at http://www.eastasiaforum.org/2011/05/14/rare-metals-after-the-japanese-nuclear-crisis/, accessed 14 May 2011.

Tse, Pui-Kwan (2011) *China's rare-earth industry* US Geological Survey).

US Department of Energy (2010) *Critical materials strategy* (Washington: US Department of Energy).

United States Geological Survey (1996) *Rare earths* (Reston: USGS).

United States Geological Survey (2004) *Rare earths* (Reston: USGS).

United States Geological Survey (2011) *Rare earths* (Reston: USGS).

Wang, Yaguang (2011) China tightens regulation of rare earth industry. *Xinhua*, 15 June.

Wū raŏn fáng zhì sī (2011) Guān yú yìn fā《 xī tǔ qǐ yè huán jìng baŏo hù hé chá gōng zuò zhǐ nán》de tōng zhī.

Xī tǔ gōng yè wū raŏn wù pái fàng biāo zhǔn (GB26451 fàng biāo) Available at http://kjs.mep.gov.cn/pv_obj_cache/pv_obj_id_41D98CE00CFE396700315824DA96D4F024250600/filename/W020110210366768105784.pdf, accessed November 2011.

Yang, Ai (2010) Japan pressed to confer before scouring seabed. *China Daily*, 27 April.

Yomiuri Shimbun (2012) Govt right to pressure China to respect WTO rules. *Yomiuri Shimbun*, 15 March.

Yonhap News (2010) Seoul to boost stockpile of rare metals by 2016. *Yonhap News*, 15 October.

Zhang, Qi (2011) New standards for rare earth sector. *China Daily*, 7 January.

Zhang, Qi and Qingfen Ding (2010a) Export tax to be raised on rare earth. *China Daily*, 15 December.

Zhōng huá rén mín gòng hé guó zhōng yāng rén mín zhèng fǔ (2011) Guó wù yuàn guān yú cù jìn xī tǔ háng yè chí xù jiàn kāng fā zhaǒn de ruò gān yì jiàn guó fā〔2011〕No. 12. Available at http://www.gov.cn/zwgk/2011-05/19/content_1866997.htm, accessed 20 December 2011.

Zhōng huá rén mín gòng hé guó shāng wù bù (2005) shāng wù bù gōng gào2005nián dì 119 hào《2006 nián xī tǔ chū koǒu qǐ yè míng dān. Zhōng huá rén mín gòng hé guó shāng wù bù. Available at http://www.mofcom.gov.cn/aarticle/b/e/200601/20060101280116.html, accessed 15 December 2010.

Zhōng huá rén mín gòng hé guó shāng wù bù (2009) Shāng wù bù gōng gào2009nián dì129hào fā bù 2010 nián xī tǔ chū koǒu qǐ yè míng dān. Zhōng huá rén mín gòng hé guó shāng wù bù.

Zhōng guó rén mín gòng hé guó guó tǔ zī yuán bù (2010a) Nán fāng wǔ shěng; xǐng(qū) shí wǔ shì xī tǔ kāi fā jiān guaǒn gōng zuò zuò tán huì tí chū xī tǔ jiān; jiàn guaǒn qū yù lián dòng jī zhì yào; yāo xiàng zòng shēn tuī jìn. Available at http://www.mlr.gov.cn/xwdt/jrxw/201011/t20101108_793032.htm, accessed 1 February 2011.

Zhōng guó rén mín gòng hé guó guó tǔ zī yuán bù (2010b) .Píng lùn: Zhōng guó xī tǔ háng yè "raǒng wài bì xiān ān nèi". Available at http://www.mlr.gov.cn/xwdt/mtsy/qtmt/201011/t20101108_792890.htm, accessed 1 February 2011.

Throwing the Baby Out with the Bathwater: Australia's New Policy on Treaty-Based Investor-State Arbitration and its Impact in Asia

LUKE NOTTAGE

Sydney Law School

Abstract: *Treaties allowing investors to initiate arbitration claims directly against host states for illegally interfering with cross-border investments are becoming increasingly common in Asia, but Australia announced in 2011 that it will no longer include such protections in future treaties. The backdrop to this decision includes keen interest from Asia in foreign direct investment (FDI) into Australia's resources sector, meaning that potential investors may not be significantly deterred by a lack of arbitration provisions in future treaties. This article argues, however, that Australia's policy shift risks undermining the entire investor-state arbitration (ISA) system, with the earliest impact being felt by major pending treaty negotiations by Australia with Japan, China and Korea (respectively); and that the shift may significantly reduce FDI flows or have other adverse effects. The article criticises the cost-benefit analysis of ISA protections in one pivotal study conducted in 2010 by an Australian Government think-tank, arguing that this assessment is insufficiently nuanced. Instead, the article presents a justification for more tailored and moderate changes to ISA provisions in future treaties. Its tentative interest-group analysis suggests, however, that there may be surprisingly few public or private constituencies that would prefer such moderate reforms, and that most may well prefer the more extreme position recently adopted by Australia, despite the damage that will be done to the ISA system as a whole. The article also argues that Australia's policy shift and think-tank analysis may make Asian countries more cautious about ISA, especially those (like the Philippines and Vietnam) which have traditionally been more cautious about this dispute resolution system.*

Introduction

Treaties in Asia increasingly allow investors from home states to bring international arbitration claims directly against host states for illegally interfering with their investments. Yet Australia decided in 2011 to no longer include such protections in future treaties, even with developing countries, which are traditionally the main targets of such protections because of their less reliable laws and court procedures. The backdrop includes keen interest from Asia in foreign direct investment (FDI) into Australia's resources sector; Australia may not need to offer such treaty protections to entice foreign investors. Yet, as outlined in the first part of this article, the policy shift risks undermining the entire investor-state arbitration (ISA) system, beginning with major pending treaty negotiations by Australia with Japan, China and Korea (respectively). This may significantly reduce FDI flows and have other adverse effects in the broader Asian region, especially in the resources sector.

The impact of Australia's policy shift depends partly on the persuasiveness of rationales given by the Government's Productivity Commission (2010) regarding trade and investment treaty policy. The second part of this article therefore critically assesses the economic theory and evidence underlying the Commission's Report. Given problems identified by that analysis as well as the many complex implications of the Government's new policy stance, the third part outlines some less radical ways for Australia – and other countries in the region – to rebalance private and public interests in the ISA system. The fourth part argues that Australia's recent experience indicates more generally that nowadays there may be surprisingly few constituencies prepared to come out strongly in favour of refining the ISA system in those moderate ways. Within many other states, there are probably more public and private interest groups now wishing to see the ISA system curtailed – along the lines recently announced by the Australian Government or, indeed, even more restrictively.

The article concludes that many other states in Asia already negotiating investment treaties with Australia are also unlikely to achieve a relaxation in its policy stance. This also significantly complicates the attempts to create the first truly regional FTA through the Trans-Pacific Partnership Agreement (TPPA) negotiations. The treaty-based ISA system, despite its remarkable expansion world-wide over the last decade and signs of growing acceptance in Asia, may well therefore end up declining significantly in the region over the medium to longer term.

Investor-State Arbitration in Asia Meets Australia's New Policy

Japan has been negotiating a Free Trade Agreement (FTA) with Australia since 2007.[1] One major sticking point has been liberalisation of Australian access to the rich Japanese market for agricultural produce, but another potential problem has also been shaping up over liberalisation and protection of Japanese investors in Australia. In 2010 more foreign direct investment (FDI) came from Japan than from the People's Republic of China, although the latter drew much more public attention. This renewed wave of Japanese investment is more diversified than the wave of investment during the 1980s, when a strong yen and asset price inflation in Japan's "bubble economy" led to large investments in tourism and real estate in Australia and other developed countries. Recent large investments from Japan are focused on securing mineral resources – in the face of strong competition from Chinese investors and some emerging interest from India and Korea – and in Australia's food and beverages sector (Drysdale, 2009).

Australia's recent political controversy over a "super profits tax" on the mining industry, which contributed to the replacement of Prime Minister Kevin Rudd with Julia Gillard (Stuart, 2010), focused the minds of some Japanese investors on the vulnerability of FDI to domestic uncertainties, even in developed countries such as Australia. One way to reduce such uncertainties, especially in large-scale ventures involving natural resources development, is to elaborate substantive protections under international law by means of FTAs (or Bilateral Investment Treaties, BITs). Like China and Korea (Bath, 2011; Kim, 2011), Japan has accelerated its treaty negotiation program over the last decade (Hamamoto, 2011). Its FTA with Indonesia signed in 2007, for example, aims at liberalised access for Japanese investors in Indonesia's resources sector (Sitaresmi, 2011). Japan also increasingly presses treaty partners to agree to allow investors to bring arbitration claims directly against host states that allegedly breach such substantive obligations – for example, by expropriation, lack of transparency or fair and equitable treatment (FET). This represents a more efficient and less politicised dispute resolution mechanism from the perspective of investors, compared to the customary international law approach (also now elaborated in treaties as an alternative) whereby the investor's home state may be mobilised to bring an inter-state claim against the host state. Japan had been seeking such investor-state arbitration (ISA) provisions in its bilateral FTA with Australia, consistently with Japan's longstanding treaty practice, notwithstanding the fact that the devastating natural disasters of 11 March 2011 and the subsequent nuclear power plant emergency have slowed the pace and shifted the dynamics of those FTA negotiations.[2]

In April 2011, however, Australia released the 'Gillard Government Trade Policy Statement' (TPS), which ran contrary to the Japanese push for ISA provisions. The Government reaffirmed that Australia's trade policy should focus first on multilateral liberalisation – and indeed, in many cases, unilateral liberalisation measures – under the World Trade Organization (WTO) system, instead of bilateral or regional deals. The difficulty with this declaration is that the WTO system provides very limited protections for investors, offering some protections for certain services sectors, but generating a more limited range of direct remedies for investors than BITs or FTAs with investment chapters. WTO treaty violations (e.g. of "national treatment" commitments) rely on an investor persuading its home state to initiate an inter-state dispute resolution process, rather than allowing investors to bring claims directly against host states (as under ISA). Furthermore, the TPS blew cold on including ISA in any future FTAs or BITs:[3]

> The Gillard Government supports the principle of national treatment – that foreign and domestic businesses are treated equally under the law. However, *the Government does not support provisions that would confer greater legal rights on foreign businesses than those available to domestic businesses.* Nor will the Government support provisions that would constrain the ability of Australian Governments to make laws on social, environmental and economic matters in circumstances where those laws do not discriminate between domestic and foreign businesses…

> *In the past, Australian Governments have sought the inclusion of investor-state dispute resolution [especially ISA] procedures in trade agreements with developing countries at the behest of Australian businesses. The Gillard Government will discontinue this practice.* If Australian businesses are concerned about sovereign risk in Australian trading partner countries, they will need to make their own assessments about whether they want to commit to investing in those countries.

These points resonated with the analysis of ISA in the Final Report of the study into Australia's Bilateral and Regional Trade Agreements commissioned by the Australian Treasurer (Productivity Commission, 2010). The Commission's Recommendation 4(c) had stated that the Gillard Government should "seek to avoid the inclusion of investor-state dispute settlement provisions in [treaties] that grant foreign investors in Australia substantive or procedural rights greater than those enjoyed by Australian investors".[4] Yet even this recommendation seemed to allow some scope for Australia to include ISA provisions in future treaties – notably, with countries with less developed legal systems. The Recommendation might have allowed Australia to cap substantive protections (e.g. against expropriation) entrenched in treaties – and underpinned by ISA rights – at the level of protection provided anyway under Australian domestic law. In effect, this would have allowed investors from Australia to take abroad – into the host-state party to such a treaty – the substantive protections entrenched by the treaty. Conversely, foreign investors into Australia from the treaty partner would not really have obtained much benefit from such a treaty regime, because they could obtain such substantive protections by suing the Australian Government in Australian courts anyway. But even such foreign investors would not have had to rely on pursuing the substantive protections through local courts in Australia; they too could instead have availed themselves of the treaty's ISA mechanism.

In fact, this sort of (quite one-sided) approach arguably characterised Australia's investment treaty practice up until the TPS. All its treaties included ISA protections *except* for the Australia–US FTA (AUSFTA, signed in 2004), the Australia New Zealand ASEAN FTA (AANZFTA, 2009) in relation only to the bilateral relationship between Australia and New Zealand, and the Investment Chapter added in February 2011 to Australia's "Closer Economic Relations" FTA with New Zealand dating back to 1982 (Mangan, 2011). Yet the omission of ISA in AUSFTA was arguably more tactical than part of a well thought out strategic plan,[5] and Australia did include ISA in FTAs with countries such as Singapore (2001) and Chile (2009) – which arguably have well-developed legal systems.

From pronouncements by Australian Government officials at public events since May 2011 (Nottage, 2013), it appears that the TPS was intended to be taken literally: Australia will no longer include ISA provisions even in treaties with developing countries.[6] Since there is far less incentive to protect one's outbound investors when negotiating with a country that has a developed legal system, this effectively means that Australia does not want ISA in any future treaties. That represents a very significant departure from Australia's practice for more than two decades. All treaties have contained ISA protections when the counterparty was a developing country, although the BIT with China only managed to incorporate limited rights to ISA as China had not yet developed into a major source of outbound investment (Eliasson, 2011).

Australia's policy shift has complex and potentially wide-ranging ramifications, especially in the Asian region (Nottage, 2011, Part I). Asian countries' investment treaties have increasingly provided for ISA protections (Bath and Nottage, 2011), paralleling the growing acceptance of international commercial arbitration as a mechanism for resolving other types of disputes (Nottage and Garnett, 2010). Yet some countries in Asia arguably remain cautious about ISA and investment liberalisation more generally (Sornarajah, 2011). For example, India and Vietnam have not acceded to the 1965 Convention on the Settlement of Investment Disputes between States and Nationals of Other States (the ICSID Convention), which is promoted by the World Bank. The ICSID Convention facilitates enforcement of ISA awards rendered against a host state party to the Convention if the proceedings are conducted in accordance with ICSID Arbitration Rules, as may be provided by the host state as one option under an investment treaty. Also noteworthy is that Asian parties appear still to be under-represented in formal ICSID or other ISA case filings. Yet part of the reason for this seems to be greater "institutional barriers" facing Asian investors or host states that may be considering ISA proceedings (Nottage and Weeramantry, 2012). Overall, the growing acceptance of ISA in the Asian region may be significantly undermined by Australia's new policy stance, if its future treaties omit ISA provisions or regional partners go on to reassess their own general approach to ISA in light of the rationales officially given for the new policy.

This article therefore takes a closer look at the official reasons and other factors that account for Australia's new-found caution about ISA, and the potential implications for the region. Unofficially, one possible reason for the TPS going beyond the Productivity Commission's Recommendation is that the Gillard Government then ruled in an alliance with the Greens, who have long been cautious about trade and investment liberalisation. Coming down hard on ISA may have lessened the blowback from the TPS's reiteration of the Government's commitment to multilateral and even unilateral liberalisation measures. The Statement's opposition to ISA may also represent a reaction – arguably, an over-reaction – to a dispute with tobacco companies over Australia's Tobacco Plain Packaging Act 2011, which was supported enthusiastically by the Greens. That legislation generated (on 27 June 2011) the first-ever notice of intent to initiate arbitration against Australia under an investment treaty, namely the 1993 BIT with Hong Kong (Nottage, 2013). Yet this can hardly be said to explain Australia's actions: none of its regional neighbours reacted to the occasional claim under ISA by eschewing all ISA in future treaties.[7]

Another important factor in Australia's case, which has also not been highlighted in recent official pronouncements on ISA policy, may have been its ongoing mining boom – a major driving force behind Australia's growing *inbound* net FDI flows, especially since 2006 (Productivity Commission, 2010, pp. 31–32). The Government may well have questioned the need to offer ISA protections to foreign investors, if they are likely to invest anyway. Yet other Asia-Pacific countries enjoying abundant natural resources – such as Brunei, Indonesia, Chile or Canada – have not resiled from ISA provisions, and no mining boom can last forever anyway.

Another general consideration behind the Gillard Government's stance may be that even without ISA in new contracts, *outbound* Australian investors retain the benefit of ISA protections under 25 of 27 existing treaties concluded since 1988, and will continue to do so until one state party terminates the treaty after expiry of its initial (or otherwise automatically renewed) term. The TPS and subsequent pronouncements at public events indicate that the Government does not intend to initiate termination of old treaties (and perhaps then renegotiation of a new treaty without the arbitration provisions), despite its new-found aversion to ISA.[8] Thus, for example, the Government has apparently "politely declined" the invitation to agree to ISA in the pending negotiations to expand the TPPA. Originally signed in 2005 by four small Asia-Pacific economies, the new TPPA plans to add an Investment Chapter and include Australia, Peru, Vietnam, the US, Canada, Mexico and possibly Japan. Yet Australia already has treaties including ISA protections vis-à-vis all the developing country partners proposed for the expanded TPPA.

The short-term risk, however, is that a country such as Vietnam will follow Australia's lead and also refuse to countenance any ISA provisions. Vietnam is already attracting extensive inbound FDI, thanks in part to its high economic growth rate, yet it presents serious political and legal risks for foreign investors (Dang, 2011). If Vietnam refuses to accept ISA provisions, this may well jeopardise TPPA negotiations in the short term. This possibility arises because partners such as Singapore and the US have been pressing for a "high-quality" regional FTA as a model or core for further regional integration initiatives, and they have included ISA in (almost) all their other treaties. In the long term, moreover, many countries – especially in Asia – may rethink their attitude towards ISA protections in light of Australia's new stance, and begin to omit them in their future bilateral treaties. A core aspect of the regional and world-wide investment treaty regime, built up slowly and painstakingly in the absence of any comprehensive multilateral investment treaty regime (within the WTO system – or elsewhere), will then start to unravel. This could well have adverse or at least complex effects on cross-border investment flows, especially regarding large-scale resource development projects in Asia.

Whether this happens will depend partly on the persuasiveness of the official reasons given by the Australian Government for its policy shift relating to ISA, as well as possible parallels between Australia's political and economic environment and that in regional neighbours.

Economic Theory and Evidence behind Australia's Policy Shift

Possible benefits of ISA

The Productivity Commission (2010, p. 269) begins its case by positing that the "principal economic rationale" for granting ISA protections to foreign investors is to overcome market failure related to foreign investment, which it concedes may improve economic output, income and social services provision. It argues first, however, that governments are unlikely to take away favourable conditions initially offered to foreign investors by expropriating their assets, because of "reputational effects" – the fear of scaring off future investors. Second, the Commission dismisses the argument that foreign investors face systematic bias compared to local investors by pointing to two studies published in the mid-2000s that suggest foreign firms in fact enjoy *advantages* compared to local competitors (Huang, 2005; Desbordes and Vauday, 2007). Yet there are good reasons to doubt the wisdom of relying heavily on these studies. Both of them analyse results from the same survey, which was conducted back in 1999–2000. That era preceded the entry into force of anti-bribery legislation in many developed countries (Burnett and Bath, 2009), and so at that time foreign (Western) firms were freer to exercise political influence. The two studies in question also do not focus on parts of the world where investment flows are of large significance for Australia, such as South and East Asia. This therefore seems a shaky empirical basis for rejecting the idea that some problems and market failures encountered by foreign firms abroad might be due to host state interference. It also contradicts many qualitative studies that highlight many of those very problems, especially for foreign investors in countries such as China (Bath, 2011), Vietnam (Dang, 2011) and Indonesia (Butt, 2011).

As for the Commission's counter-argument that reputational effects discipline host states – an assertion that is not actually substantiated by any empirical study – this may be mostly true regarding outright expropriation. Yet it is only partially true regarding indirect or "creeping" expropriation, which can be caused by government action disproportionately impacting on foreign investors, or even outright breaches of the broader FET obligation under international law. The Productivity Commission (2010, p. 268) itself provides several examples where violations have been alleged by foreign investors, and sometimes expressly upheld by arbitral tribunals (see also generally Nottage, 2013). Putative "reputational effects" were obviously insufficient to deter the behaviour in such cases. Nor did such effects seem evident in the New South Wales Government's proposal in May 2011 to renege on a promise to maintain feed-in tariff rates payable to residents who had installed solar power panels (Nottage, 2011, Part II). The Government eventually abandoned the proposal, but mainly due to opposition from householders about such retrospective legislation, rather than from panel suppliers who might have qualified as aggrieved foreign investors under investment treaties.

Returning to the economic analysis of potential benefits from ISA, as sketched by the Productivity Commission (2010, p. 269), the Commission also points to some econometric research indicating instead that adding ISA in investment treaties has "no statistically significant impact on foreign investment into that country". Yet econometric studies depend on many things, including the estimation techniques adopted. Interestingly, one method used in the study upon which the Commission relied in fact finds a highly significant relationship (at the 99 per cent confidence level) between including ISA provisions in treaties and higher inbound FDI – namely, for Regional Trade Agreements (or FTAs), albeit not for BITs alone or for FTAs and BITs combined (Berger et al., 2010, p. 17). The selection of time frame is also often important. The study cited by the Commission looks at data through to only 2004, whereas treaties with stronger forms of ISA protections (e.g. treaties concluded by China, former communist countries of Eastern Europe, or ASEAN nations) probably represent a higher proportion of all treaties concluded over the last seven years. That can presumably be correlated with very strong growth in FDI flows, at least until the dip during the global financial crisis of 2008, which might well affect the results from a regression analysis re-run with more contemporary data.

The coding, choice and measurement of variables are also crucial. The study by Berger et al. (2010) codes ISA provisions into three levels in terms of their scope, based on the important study by Jason Yackee which focused on BITs. But recent arbitral jurisprudence suggests that there are in fact four levels of protection (Eliasson, 2011). We also need to be careful not to include too many dependent variables (risking problems of auto-correlation), but not to include too few (omitting major explanatory factors).[9] In addition, even with the crucial (dependent) variable in these studies – the amount of inbound FDI – researchers face great disparities in measurement across countries, especially over lengthy time periods.

Lastly, the most important thing about econometric analysis is that it deals in aggregates. A statistically significant result at the 95 per cent confidence level for an estimated relationship between the independent and dependent variables means that we would expect it *not* to be true in around five cases out of 100. What if those cases are disproportionately clustered around countries that are already – or may probably become – of greatest significance for Australia in terms of investment flows? As a better guide for real-life policymakers, like the Australian Government contemplating treaty negotiations, econometric analysis should focus on such existing or likely partner countries – including many throughout Asia. Adopting a stance based even on overwhelmingly consistent econometric evidence of global patterns – say, to omit or severely limit ISA provisions in investment treaties – is therefore a risky strategy for Australian policymakers.

Indeed, in discussing "implications for future policy" the Productivity Commission (2010, p. 276) itself cites a Submission from the Department of Foreign Affairs and Trade (DFAT) that seems consistent with this general point:

> DFAT submitted that it already "advocates a careful, case by case approach to the inclusion of Investors State Dispute Settlement (ISDS) in Australia's international agreements", taking into account matters including the nature of the partner country's legal system, stakeholder views, precedents and the promotion [of] bilateral investment flows.

The Productivity Commission (2010, p. 270) also acknowledges that ISA provisions

> could still benefit particular investors to the extent that they shift political risks associated with investments to host governments and/or provide an avenue for compensation "after the event". In consultations following the Draft Report, it was also suggested that [ISA] could provide additional leverage to businesses when negotiating with foreign governments prior to undertaking (or during the life of) foreign investments, were the businesses willing to threaten to pursue an arbitration case against a foreign government.
>
> However, as noted in chapter 7 [of this Final Report], the Commission received no feedback from Australian businesses or industry associations indicating that ISDS provisions were of much value or importance to them. Indeed, as far as the Commission is aware, no Australian business has made use of [ISA] provisions in Australian [investment treaties], including in its [FTAs].

Yet one Australian mining company recently succeeded in a claim under the Australia–India BIT (signed in 2000)[10] and an Australian-owned mining company (with interests also in Senegal and Indonesia) is presently bringing ICSID proceedings against The Gambia, after its iron sands licence was revoked and its British Managing Director was arrested in 2008.[11] That case is admittedly under arbitration provisions contained in an investment contract, since Australia has no investment treaty with The Gambia, but it highlights one dispute among several to have arisen in that region (Askew and Ayala, 3 June 2006). Despite such risks, the new TPS precludes including ISA provisions in any future treaties with African nations.

Overall, the Commission speculates that Australian businesses did not express interest in ISA provisions during their Review because investors have found other options to be relatively attractive. For example, it suggests the possibility of negotiating specific (pre-)investment contracts including dispute resolution clauses. The Commission (2010, p. 270) does nevertheless admit that this option "is more feasible for large businesses", citing the Gorgon Gas project in Western Australia as an example. The Commission also mentions political risks insurance against expropriation. Yet such coverage is usually unavailable for other protections (such as FET) or the lengthier periods provided by investment treaties. Indeed, in a strong dissent to the majority view on ISA presented by the Productivity Commission (2010, p. 320), Associate Commissioner Andrew Stoler argued that "this is analogous to arguing against the need for a fire department because homeowners can buy property insurance".

Despite the brevity and superficiality of its analysis the Productivity Commission (2010, p. 271) reached its Finding 14.1, which underpinned its near-unequivocal Recommendation 4(c) on ISA: "There does not appear to be an underlying economic problem that necessitates the inclusion of [ISA] provisions within agreements. Available evidence does not suggest that [ISA] provisions have a significant impact on investment flows".

Risks of ISA

Already unconvinced about significant benefits from ISA, the Productivity Commission (2010, pp. 271–72) then outlines various risks involved in Australia agreeing to them in its investment treaties, highlighted also by some Submissions to its Inquiry. These include the possibility of "regulatory chill" on public authorities; the undermining of democratic (legislative and other) processes; and the disadvantaging of domestic investors, thereby distorting efficient investment flows. However, as pointed out by Stoler in his dissent to the Productivity Commission (2010, p. 320):

> Opponents of [ISA] cite cases such as where governments may back off regulating cigarette packaging due to the threat of a suit by a foreign investor. In the Associate's view, the appropriate response to these concerns is to ensure that the [ISA]-related provisions of [an investment treaty] are drafted carefully enough that they preclude challenges to those regulatory areas that Australia wants to ensure are protected (for example, health-related policies). In addition, in the Associate's view, there is reason to believe that a little bit of "regulatory chill" might be a good thing, even in Australia.

Stoler's first point suggests that concerns about an incipient claim against Australia from the tobacco industry, which eventuated in June 2011 (Nottage, 2013), may have already been quite widespread in 2010. His proposed solution – better treaty drafting – is commendable and his final point is well illustrated by Australia's recent solar panel tariffs debacle, as previously outlined (Nottage, 2011). That is, treaties backed by ISA provisions might have made the NSW Government think more carefully before announcing its drastic policy reversal in the first place. Indeed, in this case international law would have reinforced, rather than undermined, democratic values within Australian society – namely, concerns about enacting legislation with serious retrospective effects – even if there is no outright constitutional prohibition on such enactments, and the "legitimate expectations" doctrine in domestic law may be somewhat narrower than in international investment law (Nottage, 2013). Admittedly, giving foreign investors (possibly) greater substantive rights underpinned by (probably) stronger procedural rights through ISA provisions theoretically might still "crowd out" local investors. Yet in the specific instance under question, it should be recognised that Australia is woefully behind the ball in developing its solar power potential, so attracting foreign investors and suppliers in this field should provide countervailing economic benefits.

More generally, for decades the Treasury and other parts of the Government have emphasised the broader efficiencies created by allowing foreign investors to compete in the domestic market (Crotti et al., 2010), yet the Commission does not mention this broader consideration in relation to ISA policy. Perhaps the Gillard Government considers that liberalisation and growth of inbound FDI in Australia over the last two decades means that marginal efficiency gains from further investment are likely to be small and diminishing, compared to the risks involved in accepting more inbound FDI. Alternatively, the Government may be betting on the mining boom continuing into the medium term, reducing the need to offer ISA protections to attract FDI into the resources sector. Yet neither proposition has been advanced – let alone fully argued – in the TPS or other major official pronouncements.

Rather, the Productivity Commission (2010, p. 272) went on to identify concerns raised about damages awards in ISA cases, including "the degree of freedom arbitral tribunals have in determining" amounts and the "potential for large claims" by foreign investors. It also emphasised various problems identified with arbitral procedure, ranging from the lack of appeals (for substantive error of law, presumably) and putative "institutional biases and conflicts of interest, inconsistency and matters of jurisdiction, a lack of transparency and the costs incurred by participants". The Productivity Commission (2010, p. 274) therefore concludes with Finding 14.2: "Experience in other countries demonstrates that there are considerable policy and financial risks arising from ISDS provisions".

At least the Final Report abandoned the assertion contained in the Commission's Draft Report that US investors had never lost an ISA claim, after contrary data was provided in Submissions.[12] Nonetheless, it still overstates the risks of being subjected to a claim. In fact, empirical studies suggest that damages claims are much less successful in terms of awards on both liability and damages, and incur fewer (direct) costs, than conventional critiques and anecdotes tend to assume (e.g. Frank, 2011). This final point accounts for most of the rationale provided by the Commission's Report, but it still leaves us to consider the various procedural problems highlighted by the Commission. These are real, but I argue next that they can be addressed through moderate and targeted reforms to the ISA system without rejecting it in toto.

Alternative Means to Rebalance Private and Public Interests in ISA: Paths Not Taken

Compared to the Draft Report, the Productivity Commission (2010, pp. 274–76) devotes far more attention to measures for "reducing the risks" of ISA. These include:

- more precise definitions of more contentious terms, such as "expropriation" or "investment" and "most favoured nation" treatment related to ISA provisions;
- "time-limiting agreements" (e.g. where a "partner country is rapidly developing, such that its legal system can eventually resolve investment-related disputes" fairly anyway); and
- carve-outs for developed countries, as under AANZFTA – speculating that this might also be a way forward for the expanded TPPA negotiations.

The Commission also acknowledged my Submission that concerns about procedural rules in ISA

> can be reduced by the Australian Government through the inclusion of clauses in [investment treaties] that change the default rules of the ICSID or UNCITRAL. These changes could include requiring foreign investors to exhaust domestic legal channels prior to initiating arbitration, requiring that the existence of arbitration cases, documentation and awards be transparent and publicly available; and providing for arbitration appeals. One way to do so could be for Australia to develop a 'Model International Investment Agreement' that includes more tailored arbitration rules (sub. DR63, p. 1).
>
> Indeed, Australia followed this course in its agreement with Chile [2009], which contains considerably more detailed procedural requirements than for Australia's other agreements, including the requirement that investors attempt to consult with the host government prior to arbitration, the selection of arbitrators and the conduct of arbitration, as well as requiring transparency of arbitration documentation and any awards that are made.

The latter approach in fact represents only one of many possible reform strategies that had been submitted to the Commission.[13] The former approach, allowing institutions such as the Australian Centre for International Commercial Arbitration (ACICA) to develop tailored ISA rules for the Government to add in treaties as another option (Nottage and Miles, 2009), may be more flexible but have less immediate practical impact. A more intrusive procedural reform is to require "exhaustion of local remedies" in host state courts or administrative processes before allowing access to ISA, but specifying a time limit for local proceedings after which ISA can be invoked (as in many of China's investment treaties). Albeit at a cost, this would reduce – but admittedly not eliminate – the procedural advantages afforded by ISA to foreign over local investors, although foreign investors arguably often experience various comparative disadvantages when litigating in local courts.

Another reform option comprises carve-outs in treaties for various sectors, such as natural resources; or various types of measures, such as taxation measures – assisting countries like Australia interested, for example, in taxing mining companies more heavily. There can also be more broadly-worded exceptions preserving regulatory capacity, for example in relation to public health, as suggested by Associate Commissioner Stoler above. A flexible combination of these approaches, already seen in recent treaty practice world-wide (UNCTAD, 2010), seems the best way forward in terms of balancing the benefits and risks of ISA for countries such as Australia. At least on legitimacy grounds, it is arguably preferable to pressing for often one-sided obligations favouring only its investors abroad, or even (especially in the context of regional agreements) agreeing to reciprocal rights but only with partners with allegedly "developed" legal systems (as may have been envisaged by the Commission).

At least, one should consider the novel approach of allowing only inter-state dispute resolution but with the power of an investor to force its home state to initiate proceedings against the host state, as under a growing number of tax treaties – including several concluded recently by both Japan and Australia. North American investment treaties also often now provide for an interesting hybrid procedure that gives more weight to the state party while retaining some procedural rights for investors. They can initiate ISA proceedings alleging that a host state has implemented "expropriatory taxation" measures, but both states can curtail individual claims by agreeing that the measures do not constitute expropriation. This suggests a useful compromise dispute resolution process for other sensitive issues, such as public health regulations introduced by a host state (Burch et al., 2012). Unfortunately, however, the Australian Government did not examine such alternatives.

Instead, the Commission partly opened the door for the TPS to adopt a policy stance that completely eschews any form of ISA in Australia's future investment treaties. The Productivity Commission (2010, pp. 276–77) itself envisaged that investors could be left simply with the possibility of inter-state dispute resolution (not a right to activate it, as under some contemporary tax treaties: Burch et al., 2012), individually negotiated investor-state contracts or ad hoc legislative consent to arbitration (as in the Gorgon gas development), or otherwise only remedies provided by host state courts. The Commission suggested that investors may be able to obtain political risks insurance, if commercially viable. Yet such insurance is usually less extensive, especially for long-term resource developments, and often piggy-backs on government support.[14] Investors may also benefit from any general improvements in host states' legal systems thanks to overseas development assistance provided by home states, but this is typically only a long-term solution.

These non-treaty alternatives suggested by the Commission derive from an analysis that seems to have overestimated the (non-manageable) risks of ISA, while underestimating some of its general and specific benefits. It is therefore particularly worrisome that the Gillard Government in its TPS goes even further than the Commission's Recommendation, rather than adopting some more moderate and flexible approaches to addressing specific problems in the ISA system.

Private and Public Interest Group Incentives for ISA Reform

Australia's recent experience, however, suggests that few private or public interest groups are likely to have strong or unambiguous incentives to press for such a "middle way". Indeed, this may well prove to be true elsewhere, especially now that a developed country such as Australia has set off in a novel direction. This part sketches a broader "thought experiment" as to likely constituencies for ISA reform within contemporary nation states.

Private sector constituencies

Australia's private sector certainly made few submissions to the Commission's review in 2010, for example, and the Keidanren (Japan's key business federation) has not played a particularly large role in Japan's treaty practice regarding ISA (Hamamoto and Nottage, 2010). But perhaps this is unsurprising. *Large investors* can rely on informal links with host and home states to resolve cross-border disputes, and may also have the resources to take a longer-term approach. *Smaller investors* have even less knowledge of the pros and cons of ISA – thus creating a Catch-22 situation. Both types of firms are especially unlikely to be aware of the potential importance of ISA protections if their own home state has not yet been exposed to a claim.[15] Another factor may be the sectors from which the investors come, or in which they aim to invest. For example, a growing proportion of recent outbound investment from Australia also involves mining and resources (Productivity Commission, 2010, p. 33), and firms engaged in that type of business tend to have a high tolerance for various risks – making ISA protection less significant than for firms from other sectors, even if they do learn of its role.

As for *exporters*, they will generally be more interested in their government pressing for trade preferences in FTAs rather than strong investment chapters, although exporters nowadays are also increasingly investors or licensors of intellectual property (and so potentially already enjoying protections under investment chapters). *Domestic market oriented firms* are likely instead to *oppose* calls by foreign investors or their governments to "level the playing field" by allowing foreign investors to access the international arbitration process, not just idiosyncratic local courts. Even large *law firms* may be quite ambivalent, despite their greater access to the policymaking process than smaller law firms. After all, they disproportionately represent larger investors (with theoretically less need to press for ISA than small investors).[16] Admittedly, some large law firms increasingly promote ISA because growing ICSID caseloads represent potentially lucrative fees as advocates and arbitrators. It also raises law firms' profiles in the burgeoning and partly overlapping field of commercial (inter-firm) arbitration, but there remain many hundreds of investment treaties, many with ISA, and hence there remains plenty of work advising clients on how to structure investments to take advantage of such provisions. Likewise, the *academic community* appears to be split. Many specialists in international law are now appointed as consultants, experts or arbitrators in investor-state disputes. But many others are now making their mark as strong critics of the entire system.[17] As for *non-governmental organisations,* a growing voice in international rule-making, there are probably many more opposed to ISA than in favour – although those favouring it may sometimes be better funded or have significant influence on governments generally, such as pro-business groups.[18]

Public sector constituencies

Different parts of government can also be expected to adopt different views. A *foreign ministry* can be keen on treaty-based ISA because it can minimise time-consuming and expensive involvement in disputes reported by its own investors, and ISA may even avoid friction with host states. Home state investors obtain formal control over prosecuting claims themselves, so diplomats have more capacity to tell them to resolve disputes directly with their host states. If host states complain, the home state's diplomats can say that the matter is now out of their hands. Yet diplomats will probably want transparency obligations included in treaties (as in the Australia–Chile FTA) so they at least remain informed about claims lodged by their outbound investors. They can then mediate informally with the host state if necessary to maintain good diplomatic relations overall. On the other hand, foreign ministries can often have political incentives to conclude FTAs promptly for their political masters, and therefore prefer a negotiating position that minimises controversy associated with pressing for ISA provisions. Such reduced commitment to ISA seems more likely where, as in Japan (Mulgan, 2008), true negotiating authority is widely dispersed among government departments and political leaders – with multiple potential veto or blockage points.

A *justice ministry* often has the responsibility of defending international law claims brought against the government. Typically it would also provide legal opinions about the legality of governmental action, if asked beforehand. Particularly if the government takes action that generates a claim, that justice ministry might feel under threat. In any event it would need to spend time and seek resources to defend claims, which it might consider could have been allocated to more productive and less stressful pursuits. A justice ministry therefore may exhibit some reticence towards ISA, especially as it is a new and expanding field that demands careful monitoring of new developments and overall trends.

An exception to this general rule might be a justice ministry where many staff are very familiar with the legal and practical issues involved in foreign investment (e.g. because they have had experience in or coordinate closely private law firm practice) and/or they are likely to move into such legal practice (with law firms keen to retain the ex-officials' expertise in ISA proceedings in order to expand the services available to foreign investors). This pattern is found in the US Government (although the main government lawyers for its international law disputes are from the State Department), and this may be a significant factor behind the US policy of actively promoting ISA. Yet this phenomenon is probably unusual among countries world-wide, especially in the Asian region. It is not true in Australia and certainly not in Japan where government specialists in international law matters instead tend to serve within the public service for long periods. This does have the advantage of preventing a version of "regulatory capture" – biasing officials too much in favour of the ISA system because some might expect eventually to apply expertise in that field in the private sector – but the situation in countries such as Japan and Australia also means that such officials have fewer opportunities to experience how the ISA protections can help provide compensation to truly deserving foreign investors, or to constrain hasty or ill-conceived government policy measures that are likely to breach treaty obligations. Matters are unlikely to change much in the short to medium term because institutions and norms surrounding "government lawyering" tend to be path-dependent and resistant to change (e.g. Nottage and Green, 2012).

It is also possible that a justice ministry might favour ISA in a more general sense because it perceives overlaps with international commercial arbitration, which that ministry might also happen to be promoting. Yet there are significant differences between the two fields (Nottage and Miles, 2009), and different parts of the ministry may be charged with policy developments anyway. Certainly, despite its support of ACICA and arbitration law reform particularly over recent years, the Attorney-General's Department (and other parts of the Government) did not publicly voice this sort of argument to press for Australia to maintain a more proactive approach towards ISA.

There are few other parts of government that appear likely to strongly support ISA, either. A *commerce ministry*, such as the Ministry of Economy, Trade and Industry in Japan, may fall into that category. This ministry is keen to support its importers of natural resources from abroad to fuel Japan's world-class exporters of processed goods. But a commerce ministry would normally also have jurisdiction over less globally competitive industries, which it might want instead to shield from inbound FDI – by preserving "regulatory capacity", and opposing ISA for foreign investors. Other line ministries, such as a *ministry of agriculture*, are even more likely to take that approach, or (less cynically) to be concerned about treaty-based limits to their capacity for regulatory responses to emerging socioeconomic problems.

A *finance ministry* will also usually be very concerned about liability exposure from ISA claims from foreign investors. But what if this ministry, or a related entity, also has primary jurisdiction to develop policy about inbound investment (like the Department of Treasury in Australia) and to screen it in the national interest (through the Foreign Investment Review Board, FIRB, which advises the Treasurer)? What if it generally welcomes inbound FDI to promote allocative and dynamic efficiency? This should elicit a more positive view towards ISA.[19] Yet such a development seems likely to diminish as the local economy is progressively opened up to inbound FDI (perhaps reducing marginal efficiency gains) or if further economic studies begin to suggest that offering ISA does not significantly increase inbound FDI anyway (as asserted by the Commission).

Overall, therefore, this preliminary outline of both public and private sector (sub-) groups within nation-states which might potentially be interested in ISA indicates few obvious strong constituents for maintaining the present ISA system. A few commentators in Australia expressed surprise when the Commission's Final Report recommended a potentially major diminution in the likelihood of including ISA provisions in future treaties. Many more became concerned when the Gillard Government went even further with its TPS, especially as Australia's new policy seemed to be over-reacting to the notice of claim lodged by Philip Morris Asia in June 2011.[20] Yet perhaps a significant backlash against ISA should have been expected, in light of the various interest group dynamics sketched above. Policymakers in Asia should therefore consider this sort of broader backdrop and its practical implication for other parts of the region.

Conclusion

Overall, the somewhat belated emergence of ISA in Asia may well be halted by a domino effect around the region, initiated or accelerated by the Gillard Government's new policy stance. Major on-going FTA negotiations suggest that even international pressure may not create much incentive for Australia to rethink and adjust its new policy position. China, for example, will be very keen to conclude treaty negotiations with Australia to liberalise access for investments, particularly in the booming resources sector. Its recent treaty practice has been to include comprehensive ISA protections (Eliasson, 2011), which could offer particular comfort to Chinese investors in light of growing sensitivities in Australia – including some high-profile cases of FIRB blocking inbound FDI (Bath, 2011). Yet China may consider that omitting full-scale ISA may result in the Australian Government becoming more willing to let in Chinese FDI on a case-by-case basis – after all, if the Government subsequently interferes with investments from China, there exists reduced claim potential if ISA is omitted. The Chinese Government may also not lose much "face" by going along with Australia's new stance, because earlier Chinese treaties had omitted or limited ISA provisions too.

Like China and several other Asian countries, Japan is also desperate to invest in and secure natural resources from Australia, especially in the energy sector following the "3-11" disasters in June – which threw the nuclear power industry into disarray. Already, Japan has made an exception in its treaty practice by omitting ISA at the request of the Philippines, in the FTA signed in 2006 (Hamamoto and Nottage, 2010). This precedent should make its omission easier in the FTA being negotiated with Australia, which anyway has a more developed legal system, especially as Japanese investors have a much longer and very positive experience of operating in Australia. In addition, really for the first time, the media in Japan have recently started to question the merits of ISA provisions more generally. Some commentators point to Korea, where the main opposition party objected to ratification of the FTA signed in 2007 with the US (and ratified there this year), partly because of fears that US investors would disproportionately invoke ISA provisions contained in that treaty (Borowiec, 2011). A concern was that a similar situation might arise under the expanded TPPA, if Japan joined those negotiations (as became more likely from November 2011) and if ISA were to be included. Korea, too, in its FTA negotiations with Australia, may now back away from its longstanding policy of including ISA protections (outlined by Kim, 2011).

Thus, for various reasons but primarily to secure other investment and trade benefits from bilateral FTA negotiations pending with Australia, even these three major economies in Asia may break with their usual treaty practice nowadays. They may instead go along with the TPS stance of omitting ISA in their respective FTA negotiations with Australia. This would probably have ripple-on effects. For example, TPPA partners other than Japan and Australia – such as Vietnam or Malaysia – would be more likely to espouse reasons like those given by the Gillard Government or the Productivity Commission (2010) to press for omission of ISA. Even the US may accede to such a request in order to secure its first regional FTA in the Asia-Pacific region for geopolitical reasons – to counteract China's active "FTA diplomacy" in Asia over the last decade – as much as for economic reasons. Like the Gillard Government, arguably the Obama Administration may decide that it can partially appease leftist domestic opponents of ISA by omitting it in the TPPA. It may still satisfy US business interests by pointing to other benefits from that FTA – including expanded exports into Asia, to revive the lethargic American economy.

Perhaps such a brave new world, free of ISA rights, does have some benefits. Perhaps ISA provisions did not really contribute much to higher and sustainable investment flows – or they did but increasingly risked becoming counterproductive, now that many host countries have already benefited from expanded inbound FDI and are improving their domestic legal systems. Yet developing countries, in particular, still lack investment capital and often struggle to entrench the rule of law (Bath and Nottage, 2011). Resiling from ISA is more likely to be beneficial in the short term for countries such as Australia that are enjoying an exceptional mining boom, and which maintain protections for foreign investors thanks to existing treaties as well as a more developed legal system. Yet these conditions may well change over the long term, and meanwhile an aversion to ISA may spread among very different countries – particularly in Asia. Rebuilding the treaty-based ISA system will then prove difficult or impossible. A domino effect from Australia's new policy stance would certainly undermine the "bottom-up" or "step-by-step" approach towards developing a harmonised framework for protecting cross-border investment, which had been slowly emerging – in Asia and other parts of the world – after "top-down" multilateral initiatives foundered a decade ago.

The global financial crisis in 2008 and the Eurozone crisis in 2011 have certainly raised the broader question of whether any treaty-based framework can ever really secure regional or global socioeconomic stability. Yet a reversion to ad hoc inter-state political solutions seems unlikely to be conducive to economic activity, and risks either needlessly escalating disputes or dampening legitimate claims by investors against host states. A better approach is to rebalance legal and political mechanisms to meet contemporary needs and expectations (Nottage, 2009). Many reform options exist, including several in relation to treaty-based ISA (Burch et al., 2012). Australia and Asia should not throw out the baby with the bathwater.

Acknowledgments

This article draws on research for the project on 'Fostering a Common Culture in Cross-Border Dispute Resolution: Australia, Japan and the Asia-Pacific', supported by the Commonwealth through the Australia-Japan Foundation which is part of Australia's Department of Foreign Affairs and Trade. For helpful feedback on earlier versions of this article, thanks are due to anonymous reviewers as well as Shiro Armstrong, Kate Barclay, Peter Drysdale, Graeme Smith and other participants in the symposium on 'International Politics of Resources: China, Japan and Korea's Demand for Energy, Minerals and Food' held over 28–29 July 2012 at the University of Technology Sydney.

Notes

1. See http://dfat.gov.au/FTA/ajFTA/index.html.
2. See http://www.eastasiaforum.org/2011/07/30/Japan-s-3-11-disaster-and-the-FTA-negotiations-with-Australia/.
3. See http://www.dfat.gov.au/publications/trade/trading-our-way-to-more-jobs-and-prosperity.html (emphasis added).
4. See Productivity Commission, 2010, p. 285. Available at http://www.pc.gov.au/projects/study/trade-agreements.
5. Although it was subsequently justified due to Australia's confidence in America's well-developed legal system, Australian negotiators around 2003–04 would not have been unaware that Australia had traditionally been a net capital importer (thus opening itself up to more potential arbitration claims, from future US investors), while the US itself had been reassessing its own stance relating

to ISA in light of some recent claims brought by Canada under the North American Free Trade Agreement. See Nottage and Miles, 2009.

6. Conceivably, the Gillard Government could have argued that its Trade Policy Statement simply takes the Productivity Commission's Recommendation to a logical conclusion. That is, because ISA almost always gives foreign investors "greater procedural rights" than local investors (limited to local court proceedings against governmental interference with investment rights), the Commission was suggesting that Australia should eschew ISA completely. The Government has not, however, justified its stance in this way. Rather, it has acknowledged going beyond the Commission's Recommendation, emphasising Australia's preference for "non-discrimination" (cf Nottage, 2013). In any case, the Commission did not seem to intend to advocate abandoning ISA altogether; otherwise, it would have recommended precisely that.
7. Most such claims have also been brought (always against central governments) by well-resourced foreign investors, according to case filings at http://icsid.worldbank.org/ICSID/. See, for example, *Chevron v Bangladesh* (ICSID Case ARB/06/10), *Alstom Power v Mongolia* (ARB/04/10), *Impregilo v Pakistan* (ARB/03/3), *Fraport v The Philippines* (ARB/03/25), *SGS v Pakistan* (ARB/01/13) and *Mobil Oil v New Zealand* (ARB/87/2).
8. The ostensible reason is that such termination could prejudice the rights of existing investors. Yet Australia's treaties routinely provide that protections endure for 10–20 years after (unilateral) termination anyway; and states can also agree to terminate or revise treaties on agreed terms. See Nottage, 2013.
9. Compare, for example, the extra variables used in Crotti et al., 2010: this study then found significant effects on inbound FDI into Australia particularly from FTAs with investment chapters (albeit without differentiating between the levels of ISA protections contained in them).
10. See Robertson, 2012. On India's active investment treaty program nowadays, especially as it emerges as a major source of outbound FDI, see generally Ranjan, 2011.
11. *Carnegie Minerals (Gambia) Limited v Republic of The Gambia* (ICSID Case No. ARB/09/19) at http://icsid.worldbank.org/ICSID/.
12. See Submission No. DR62 (from Mark Kantor), reiterated in Submission No. DR63 (from myself), available at http://www.pc.gov.au/projects/study/trade-agreements/submissions; and http://www.eastasiaforum.org/2010/09/08/Australian-versus-Japanese-approaches-towards-investor-state-arbitration/. Unfortunately, following its usual practice, the Draft Report (July 2010) is no longer available on the Commission's website.
13. For a summary of these and more extensive alternative reform options, including major disadvantages associated with each, see Appendix A in Nottage, 2011 (available at http://ssrn.com/abstract=1860505).
14. Political risks insurance, both formal and informal (mobilising the home state), may be particularly problematic for small- and medium-sized investors, with less financial and political bargaining power. Conversely, such firms appear to be significant users of ISA: OECD, 2012, pp. 16–17.
15. This is likely to have been one factor behind the lack of Australian business sector input into the Commission's Inquiry over 2010. Peak business interest groups have become much more interested in ISA protections since the Philip Morris Asia claim against Australia was widely reported from mid-2011 (as outlined in Nottage, 2013). For an example of a large Australian firm that seemed capable of indirectly exercising diplomatic leverage to extract itself from a resources investment in Africa, see Askew and Ayala, 2006.
16. The picture is further complicated in countries such as Australia, as large law firms have generated an increasing proportion of fee income from litigation and other legal services provided to the Government since deregulation in 1999; see Nottage and Green, 2011. If foreign investors sought to retain such firms when claiming against the Australian Government, they might be reluctant to take on such cases as they might have a conflict of interest or it might jeopardise other potential work for the Government.
17. See, for example, the 'Public Statement on the International Investment Regime' (31 August 2010), available at http://www.osgoode.yorku.ca/public_statement/.
18. Consider the invitation-only workshop on ISA policy convened by the Commission on 29 September 2010, in response to various Submissions including some (like mine) critical of its Draft Report. It involved a few other academics (economists including Dr Emma Aisbett, one of the abovementioned 'Public Statement' signatories), some officials, and representatives from AFTINET

(the Australian Fair Trade and Investment Network) and the Australian Council of Trade Unions. The business sector was not represented. See the participant list in Productivity Commission, 2010, p. 336.

19. See http://www.eastasiaforum.org/2009/07/24/China-national-security-and-investment-treaties/.
20. See, for example, *The ACICA News*, September 2011, especially pp. 10–19, available at http://acica.org.au/assets/media/news/TheACICANewsSeptember2011.pdf.

References

Askew, Kate and Violeta Ayala (2011) Slick operator. *Sydney Morning Herald*, 3 June. Available at http://www.smh.com.au/articles/2006/2006/2002/1148956541283.html, accessed 19 March 2013.

Bath, Vivienne (2011) The quandary for Chinese regulators: Controlling the flow of investment into and out of China, in Vivienne Bath and Luke Nottage (eds), *Foreign investment and dispute resolution law and practice in Asia*, pp. 68–89 (London: Routledge).

Bath, Vivienne and Luke Nottage (2011) Foreign investment and dispute resolution law and practice in Asia: An overview, in Vivienne Bath and Luke Nottage (eds), *Foreign investment and dispute resolution law and practice in Asia*, pp. 1–24 (London: Routledge).

Berger, Axel, Matthias Busse et al. (2010) Do trade and investment agreements lead to more FDI? Accounting for key provisions inside the black box. WTO Staff Working Paper ERSD-2010-13. Available at http://www.WTO.org/english/res_e/reser_e/ersd201013_e.htm, accessed 19 March 2013.

Borowiec, Steven (2011) The gloves are off in Korea's FTA debate. *Asia Times*, 5 November. Available at http://www.atimes.com/atimes/Korea/MK05Dg01.html, accessed 19 March 2013.

Burch, Micah, Luke Nottage et al. (2012) Appropriate treaty-based dispute resolution for Asia-Pacific commerce in the 21st century. *UNSW Law Journal* 35(3), pp. 1013–43.

Burnett, Robin and Vivienne Bath (2009) *The law of international business in Australasia* (Sydney: Federation Press).

Butt, Simon (2011) Foreign investment in Indonesia: The problem of legal uncertainty, in Vivienne Bath and Luke Nottage (eds), *Foreign investment and dispute resolution law and practice in Asia*, p. 112–34 (London: Routledge).

Crotti, Simon, Tony Cavoli et al. (2010) The impact of trade and investment agreements on Australia's inward FDI flows. *Australian Economic Papers*, pp. 259–75.

Hop, Dang (2011) Legal issues in Vietnam's FDI law: Protections under domestic law, bilateral investment treaties and sovereign guarantees, in Vivienne Bath and Luke Nottage (eds), *Foreign investment and dispute resolution law and practice in Asia*, pp. 225–41 (London: Routledge).

Desbordes, Rodolphe and Julien Vauday (2007) The political influence of foreign firms in developing countries. *Economics & Politics* 19(3), pp. 421–51.

Drysdale, Peter (2009) *Australia and Japan: A new economic partnership in Asia* (Canberra: Austrade).

Eliasson, Nils (2011) China's investment treaties: A procedural perspective, in Vivienne Bath and Luke Nottage (eds), *Foreign investment and dispute resolution law and practice in Asia*, pp. 90–111 (London: Routledge).

Frank, Susan (2011) Rationalizing costs in investment treaty arbitration. *Washington University Law Review* 88, pp. 769–88.

Hamamoto, Shotaro (2011) A passive player in international investment law: Typically Japanese?, in Vivienne Bath and Luke Nottage (eds), *Foreign investment and dispute resolution law and practice in Asia*, pp. 53–67 (London: Routledge).

Hamamoto, Shotaro and Luke Nottage (2010) Foreign investment in and out of Japan: Economic backdrop, domestic law, and international treaty-based investor-state dispute resolution. *Sydney Law School Research Paper* 10(145). Available at http://ssrn.com/abstract=1724999, accessed 19 March 2013.

Huang, Yasheng (2005) Are foreign firms privileged by their host governments? Evidence from the 2000 world business environment survey. MIT Sloan School of Management Working Paper 4538-04. Available at http://ssrn.com/abstract=72122, accessed 19 March 2013.

Kim, Joongi (2011) The evolution of Korea's investment treaties and investor-state dispute settlement provisions, in Vivienne Bath and Luke Nottage (eds), *Foreign investment and dispute resolution law and practice in Asia*, pp. 211–24 (London: Routledge).

Mangan, Mark (2010) Australia's investment treaty program and investor-state arbitration, in Luke Nottage and Richard Garnett (eds), *International arbitration in Australia*, pp. 191–221 (Sydney: Federation Press).

Mulgan, Aurelia George (2008) Japan's FTA politics and the problem of agricultural trade liberalisation. *Australian Journal of International Affairs* 62(2), pp. 164–78.

Nottage, Luke (2009) Asia-Pacific regional architecture and consumer product safety regulation for a post-FTA era. Sydney Law School Research Paper 09(125). Available at http://ssrn.com/abstract=1509810, accessed 19 March 2013.

Nottage, Luke (2011) The rise and possible fall of investor-state arbitration in Asia: A skeptic's view of Australia's 'Gillard Government Trade Policy Statement'. Sydney Law School Research Paper 11(32). Available at http://ssrn.com/abstract=1860505, accessed 19 March 2013.

Nottage, Luke (2013) Consumer product safety regulation and investor-state arbitration policy and practice after *Philip Morris Asia v Australia*, in Leon Trakman et al. (eds), *Regionalism in international investment law*, p. 452–74 (Oxford: Oxford University Press).

Nottage, Luke and Richard Garnett (2010) Introduction, in Luke Nottage and Richard Garnett (eds), *International arbitration in Australia*, pp. 1–21 (Sydney: Federation Press).

Nottage, Luke and Stephen Green (2012) Who defends Japan?: Government lawyers and justice system reform in Japan. *Asian-Pacific Law and Policy Journal* 13(1), pp. 129–73.

Nottage, Luke and Kate Miles (2009) "Back to the future" for investor-state arbitrations: Revising rules in Australia and Japan to meet public interests. *Journal of International Arbitration* 25(1), pp. 25–58.

Nottage, Luke and J. Romesh Weeramantry (2012) Investment arbitration in Asia: Five perspectives on law and practice. *Arbitration International* 28(1), pp. 19–62.

OECD (Organization for Economic Cooperation and Development) (2012) Investor-state dispute settlement (Paris, 16 May).

Pekkanen, Saadia M. (2008) Japan's aggressive legalism: Law and foreign trade politics beyond the WTO (Stanford: Stanford University Press).

Productivity Commission (2010) Bilateral and regional trade agreements: Research report (Canberra, 13 December).

Ranjan, Prabhash (2011) The "object and purpose" of Indian international investment agreements: Failing to balance investment protection and regulatory power, in Vivienne Bath and Luke Nottage (eds), *Foreign investment and dispute resolution law and practice in Asia*, pp. 192–210 (London: Routledge).

Robertson, Donald (2012) *Protecting your investments in foreign courts*, 6 March. Available at http://www.freehills.com/7902.aspx, accessed 19 March 2013.

Sitaresmi, Sita (2011) The Japan-Indonesia economic partnership agreement: An energy security perspective, in Vivienne Bath and Luke Nottage (eds), *Foreign investment and dispute resolution law and practice in Asia*, pp. 135–52 (London: Routledge).

Sornarajah, Muthucumaraswamy (2011) Review of Asian views on foreign investment law, in Vivienne Bath and Luke Nottage (eds), *Foreign investment and dispute resolution law and practice in Asia*, pp. 242–54 (London: Routledge).

Stuart, Nicholas (2010) Rudd's way: November 2007 – June 2010 (Carlton: Scribe Publications).

UNCTAD (United Nations Conference on Trade and Development) (2010) World investment report. Available at http://www.unctad.org/templates/WebFlyer.asp?intItemID=5539 & lang=5531, accessed 19 March 2013.

Voeten, Erik (2010) Regional judicial institutions and economic cooperation: Lessons for Asia? ADB Working Paper Series on Regional Economic Integration 65. Available at http://ideas.repec.org/s/ris/adbrei.html, accessed 19 March 2013.

Index

Note: Page numbers in **bold** type refer to **figures**
Page numbers in *italic* type refer to *tables*
Page numbers followed by 'n' refer to notes